DATE DUE

DEMCO 38-297

John Stillwell

Geometry of
Surfaces

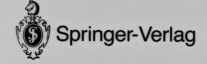
Springer-Verlag

universitext

Universitext

Universitext

Editors (North America): J.H. Ewing, F.W. Gehring, and P.R. Halmos

(continued after index)

John Stillwell

Geometry of Surfaces

With 165 Figures

Springer-Verlag
New York Berlin Heidelberg London Paris
Tokyo Hong Kong Barcelona Budapest

John Stillwell
Mathematics Department
Monash University
Clayton, Victoria 3168
Australia

Mathematics Subject Classifications: 51-01, 54-01

Library of Congress Cataloging-in-Publication Data
Stillwell, John.
   Geometry of surfaces/John Stillwell.
      p.   cm. — (Universitext)
   Includes bibliographical references and index.
   ISBN 0-387-97743-0 (Springer-Verlag New York Berlin Heidelberg:
acid-free paper). — ISBN 3-540-97743-0 (Springer-Verlag Berlin
Heidelberg New York: acid-free paper)
   1. Surfaces of constant curvature.   I. Title.   II. Series.
QA645.S75   1992
516.3'62 — dc20   91-36341

Printed on acid-free paper.

Production managed by Francine Sikorski; manufacturing supervised by Robert Paella.
Photocomposed copy prepared from the author's ChiWriter file using LaTeX.
Printed and bound by R.R. Donnelley & Sons, Harrisonburg, VA.
Printed in the United States of America.

9 8 7 6 5 4 3 2 1

ISBN 0-387-97743-0 Springer-Verlag New York Berlin Heidelberg
ISBN 3-540-97743-0 Springer-Verlag Berlin Heidelberg New York

To Wilhelm Magnus
In Memoriam

Preface

Geometry used to be the basis of a mathematical education; today it is not even a standard undergraduate topic. Much as I deplore this situation, I welcome the opportunity to make a fresh start. Classical geometry is no longer an adequate basis for mathematics or physics—both of which are becoming increasingly geometric—and geometry can no longer be divorced from algebra, topology, and analysis. Students need a geometry of greater scope, and the fact that there is no room for geometry in the curriculum until the third or fourth year at least allows us to assume some mathematical background.

What geometry should be taught? I believe that the geometry of surfaces of constant curvature is an ideal choice, for the following reasons:

1. It is basically simple and traditional. We are not forgetting euclidean geometry but extending it enough to be interesting and useful. The extensions offer the simplest possible introduction to fundamentals of modern geometry: curvature, group actions, and covering spaces.

2. The prerequisites are modest and standard. A little linear algebra (mostly 2×2 matrices), calculus as far as hyperbolic functions, basic group theory (subgroups and cosets), and basic topology (open, closed, and compact sets).

3. (Most important.) The theory of surfaces of constant curvature has maximal connectivity with the rest of mathematics. Such surfaces model the variants of euclidean geometry obtained by changing the parallel axiom; they are also projective geometries, Riemann surfaces, and complex algebraic curves. They realize all of the topological types of compact two-dimensional manifolds. Historically, they are the source of the main concepts of complex analysis, differential geometry, topology, and combinatorial group theory. (They are also the source of some hot research topics of the moment, such as fractal geometry and string theory.)

The only problem with such a deep and broad topic is that it cannot be covered completely by a book of this size. Since, however, this is the size of book I wish to write, I have tried to extend my formal coverage in two ways: by exercises and by informal discussions. The exercises include many important and interesting results that could not be fitted into the main text. Where these results are difficult, they have been broken into steps, which I hope are of manageable size. There is an informal discussion at the end of each chapter, sketching historical background, related mathematical ideas, other viewpoints, and generalizations.

Because of their deliberate informality, those discussions may mean different things to different readers, but I hope that they give at least a glimpse of a world vastly larger than the one I have covered formally. Sufficient references are given to enable intrepid readers to explore this world on their own. (References are given in the author [year] format and are collected at the end of the book.) Four references deserve particular mention, as they help to complement my approach. Nikulin and Shafarevich [1987] also extends euclidean geometry by group actions. It goes further in the direction of the third dimension, but not as far into spherical and hyperbolic geometry, or topology because it relies on strictly elementary methods. Jones and Singerman [1987] gives a geometric view of complex analysis, obtaining analytically some of the results we obtain by algebraic and geometric methods. Magnus [1974] and Gray [1986] salvage the historical treasures of the subject, Magnus from group theory and Gray from the theory of differential and algebraic equations.

This book owes a lot to my late teacher, Carl Moppert, who introduced me to the elegant reflection arguments that are common to euclidean, spherical, and hyperbolic geometry. Moppert was a student of Ostrowski, who was a student of Klein, so it may be hoped that some of Klein's spirit lives in the present book. My own students Paul Candler, Matthew Drummond, Greg Findlow, Chris Hough, Thomas Lumley, Monica Mangold, Tony Mason, Jane Paterson, Dawn Tse, and Axel Wabenhorst have also made important contributions in their turn, finding many mistakes and suggesting improvements. Extra thanks are due to Abe Shenitzer, who made a thorough critique of the manuscript just before final revision, and to Anne-Marie Vandenberg and Gertrude Nayak for typing.

I also hope to transmit some of the spirit of Magnus, and his teacher Dehn. It was Dehn who first noticed the remarkable combinatorial properties of non-euclidean tessellations and saw how they simultaneously solve problems in group theory and topology. Wilhelm Magnus not only made great contributions to group theory himself, but also brought the geometric methods to a wider audience. His death just as this book was being completed was sad for all who knew him, but happily his ideas live on and are growing in vigor.

Clayton, Victoria, Australia   John Stillwell

Contents

1

The Euclidean Plane

1.1   Approaches to Euclidean Geometry

The subject of this chapter, the euclidean plane, can be approached in many ways. One can take the view that plane geometry is about points, lines, and circles, and proceed from "self-evident" properties of these figures (axioms) to deduce the less obvious properties as theorems. This was the classical approach to geometry, also known as *synthetic*. It was based on the conviction that geometry describes actual space and, in particular, that the theory of lines and circles describes what one can do with ruler and compass. To develop this theory systematically, Euclid (c. 300 BC) stated certain plausible properties of lines and circles as axioms and derived theorems from them by pure logic. Actually he occasionally made use of unstated axioms; nevertheless his approach is feasible and it was eventually made rigorous by Hilbert [1899].

Euclid's approach has some undeniable advantages. Above all, it presents geometry in a pure and self-contained manner, without use of "nongeometric" concepts. One feels that the "real reasons" for geometric theorems are revealed in such a system. Visual intuition not only supplies the axioms, it can also prompt the steps in a proof—the choice of construction lines, for example—so that some extremely short and elegant proofs result.

Nevertheless, with the enormous growth of mathematics over the last two centuries, Euclid's approach has become isolated and inefficient. It is isolated because euclidean geometry is no longer "the" geometry of space and the basis for most of mathematics. Nowadays, numbers and sets are regarded as more fundamental than points and lines. They form a much broader basis, not only for geometry, but for mathematics as a whole. Moreover, geometry can be built more efficiently on this basis because the powerful techniques of algebra and analysis can be brought into play.

The construction of geometry from numbers and sets is implicit in the coordinate geometry of Descartes, though Descartes, in fact, took the classical view that points, lines, and curves had a prior existence, and he regarded coordinates and equations as merely a convenient way to study them. Perhaps the first to grasp the deeper value of the coordinate approach was Riemann [1854], who wrote the following:

It is well known that geometry assumes as given not only the concept of space, but also the basic principles of construction in space. It gives only nominal definitions of these things; their determination being in the form of axioms. As a result, the relationships between these assumptions are left in the dark; one does not see whether, or to what extent, connections between them are necessary, or even whether they are *a priori* possible.

Riemann went on to outline a very general approach to geometry in which "points" in an "n-dimensional space" are n-tuples of numbers, and all geometric relations are determined by a *metric* on this space, a differentiable function giving the "distance" between any two "points". This *analytic* approach allows a vast range of spaces to be considered simultaneously, and Riemann found that their geometric properties were largely controlled by a property of the metric he called its *curvature*.

The concept of curvature illuminates the axioms of euclidean geometry by showing them to hold only in the presence of *zero* curvature. In particular, the euclidean plane is a two-dimensional space of zero curvature (though not the only one, as we shall see in Chapter 2). It also becomes obvious what the natural alternatives to euclidean geometry are—those of constant positive and negative curvature—and one can pinpoint precisely where change of curvature causes a change in axioms.

Riemann set up analytic machinery to study spaces whose curvature varies from point to point. However, simpler machinery suffices for spaces, and especially surfaces, of constant curvature. The reason is that the geometry of these spaces is reflected in *isometries* (distance-preserving maps) and isometries turn out to be easily understood. This approach is due to Klein [1872]. The concept of isometry actually fills a gap in Euclid's approach to geometry, where the idea of "moving" one figure until it coincides with another is used without being formally recognized. Thus, when geometry is based on coordinates *and* isometries, it is possible to enjoy the benefits of both the analytic and synthetic approaches.

We shall assume that the reader knows how the basic notions of point, line, length, circle, and angle are handled in analytic geometry. Then our first task is to define and investigate the notion of isometry.

1.2   Isometries

The *euclidean plane* is the set
$$\mathbb{R}^2 = \{(x, y) \mid x, y \in \mathbb{R}\}$$
together with the *euclidean distance* $d(P_1, P_2)$ between points $P_1 = (x_1, y_1)$ and $P_2 = (x_2, y_2)$ given by
$$d(P_1, P_2) = \sqrt{(x_2 - x_1)^2 + (y_2 - y_1)^2}.$$

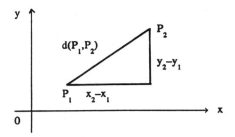

FIGURE 1.1.

One could call d the "Pythagorean distance" because it represents the length of the hypotenuse $P_1 P_2$ of a right-angled triangle with sides $x_2 - x_1$, $y_2 - y_1$ according to Pythagoras's theorem (Figure 1.1). Since it is rare to consider other distance functions on $\mathbb{R}^2$, we often denote the euclidean plane simply by $\mathbb{R}^2$, the euclidean distance function being understood.

A *euclidean isometry*, or *isometry of* $\mathbb{R}^2$, is a function $f : \mathbb{R}^2 \to \mathbb{R}^2$ which preserves euclidean distance, that is,

$$d(f(P_1), f(P_2)) = d(P_1, P_2) \quad \text{for all } P_1, P_2 \in \mathbb{R}^2.$$

The concept of isometry captures Euclid's idea of "motion" of geometric figures, except that an isometry "moves" the whole plane rather than a single figure. This enables us to make sense of the "product" of isometries and hence bring group theory into play. Admittedly it is not yet clear that isometries form a group; this will emerge as we later determine isometries explicitly. (The unclear part is the existence of inverses. Each isometry is one-to-one—because points at nonzero distance cannot have images at zero distance—but it is not clear that each isometry is onto.)

If f and g are isometries, then their *product* fg is the isometry h defined for all $P \in \mathbb{R}^2$ by $h(P) = f(g(P))$. Thus, fg is "substitution of g in f" (and *not* apply f then g" which is gf). This agrees with the usual convention of writing functions on the left. It also agrees with the order for matrix products, as we shall see when we later represent isometries by matrices, and with the order of products in the fundamental group, as we shall see when we interpret isometries topologically in Chapter 6.

The most fundamental examples of isometries follow. In each example, f is defined by giving the coordinates x', y' of $f(P)$ in terms of the coordinates x, y of P.

Example 1. *Translation* $t_{(\alpha,\beta)}$ *of O to* (α, β)

$$x' = \alpha + x,$$
$$y' = \beta + y.$$

Example 2. *Reflection in the x-axis*

$$x' = x,$$
$$y' = -y.$$

It is obvious that Examples 1 and 2 leave the square of the distance

$$(x_2 - x_1)^2 + (y_2 - y_1)^2$$

invariant. Hence, they are isometries. Less obvious is the invariance of distance in

Example 3. *Rotation r_θ about O through angle θ.*

$$x' = x \cos \theta - y \sin \theta,$$
$$y' = x \sin \theta + y \cos \theta.$$

The square of the distance becomes

$$(x_2' - x_1')^2 + (y_2' - y_1')^2 = (x_2 \cos \theta - y_2 \sin \theta - x_1 \cos \theta + y_1 \sin \theta)^2$$
$$+ (x_2 \sin \theta + y_2 \cos \theta - x_1 \sin \theta - y_1 \cos \theta)^2,$$

in which all the terms $x_i y_j$ cancel out, and all the other terms have coefficient $\cos^2 \theta + \sin^2 \theta = 1$, giving

$$(x_2' - x_1')^2 + (y_2' - y_1')^2 = x_2^2 - 2x_2 x_1 + x_1^2 + y_2^2 - 2y_2 y_1 + y_1^2$$
$$= (x_2 - x_1)^2 + (y_2 - y_1)^2,$$

as required.

The isometries in Examples 1 and 2 obviously have inverses, which are also isometries. In fact, $t_{(\alpha,\beta)}^{-1} = t_{(-\alpha,-\beta)}$ and $\bar{r}^{-1} = \bar{r}$. It can also be checked (see Exercises) that $r_\theta^{-1} = r_{-\theta}$. Since the product of isometries is associative and the identity function is an isometry, it follows that the isometries generated by the $t_{(\alpha,\beta)}$, $\bar{r}$, and r_θ form a group under substitution. Over the next two sections we shall show that *all* isometries of $\mathbb{R}^2$ are generated by the $t_{(\alpha,\beta)}$, $\bar{r}$, and r_θ, hence *the isometries of $\mathbb{R}^2$ form a group*. Since the euclidean geometric properties of $\mathbb{R}^2$ are precisely those properties invariant under isometries, this brings us to Klein's view of euclidean geometry—the study of properties invariant under the euclidean isometry group.

Since we are assuming the basics of euclidean geometry, our point of view will be somewhat different. We wish to study euclidean isometries to anticipate properties of *non*-euclidean isometries (spherical and hyperbolic) which will appear later. It turns out that in all cases a key role is played by rotations and reflections; hence we shall now look more closely at these in the euclidean case.

Exercises

1.2.1. Show that $r_\theta r_\phi = r_{\theta+\phi}$ and hence that $r_\theta^{-1} = r_{-\theta}$.

1.2.2. Show that any line $\lambda x + \mu y + \nu = 0$ is the set of points equidistant from two suitably chosen points (γ_1, δ_1) and (γ_2, δ_2). Hence, conclude (without assuming all isometries are generated by the $t_{(\alpha,\beta)}$, $\bar{r}$, and r_θ) that all isometries map lines to lines.

1.3   Rotations and Reflections

There is nothing special about the origin as a point of rotation; in fact there is nothing special about the origin at all. This is because an arbitrary point (α, β) is carried to O by the isometry $t_{(\alpha,\beta)}^{-1}$; hence (α, β) and O have exactly the same geometric properties. In particular, we can rotate $\mathbb{R}^2$ about (α, β) through angle θ by translating (α, β) to O, rotating about O, then translating O back to (α, β). Thus *rotation about (α, β) through angle θ exists* and is the isometry $t_{(\alpha,\beta)} r_\theta t_{(\alpha,\beta)}^{-1}$.

In algebraic language, $t_{(\alpha,\beta)} r_\theta t_{(\alpha,\beta)}^{-1}$ is called the *conjugate* of r_θ by $t_{(\alpha,\beta)}$. Conjugation captures the idea of "doing the same thing but at a different place," or of reexpressing an operation in a new coordinate system [in this case choosing (α, β) to be a new origin]. Just as we can obtain rotation about (α, β) by conjugating rotation r_θ about O with an isometry which carries O to (α, β), we can obtain reflection in any line L by conjugating $\bar{r}$ with an isometry f which carries the x-axis to L.

If L passes through O, it is enough to take f to be the rotation r_ϕ that carries the x-axis to L. If not, let $f = r_\phi t$, where t is a translation which brings O onto L. If L crosses the x-axis, at $x = \gamma$, say, then $t = t_{(\gamma,0)}$. If L crosses the y-axis, at $y = \delta$, then $t = t_{(0,\delta)}$.

Reflections are the most fundamental isometries because all isometries are products of them. We shall show in the next section that any isometry of $\mathbb{R}^2$ is the product of one, two, or three reflections. But first we shall show that any translation or rotation is the product of two.

Theorem. *Any translation or rotation is the product of two reflections. Conversely, the product of two reflections is a rotation or translation.*

Proof. By suitable choice of the y-axis (namely, as a line parallel to the direction of translation) we can assume that the given translation is $t_{(0,\delta)}$. I claim that $t_{(0,\delta)}$ is the product of the reflection $\bar{r}$ in the line $y = 0$ and the reflection $t_{(0,\delta/2)} \bar{r} t_{(0,\delta/2)}^{-1}$ in the line $y = \delta/2$. This is intuitively obvious (Figure 1.2), and is confirmed by calculation because

$$(x, y) \mapsto (x, -y) \quad \text{by } \bar{r}$$
$$\mapsto (x, -y - \delta/2) \quad \text{by } t_{(0,\delta/2)}^{-1}$$

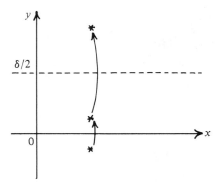

FIGURE 1.2.

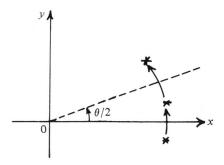

FIGURE 1.3.

$$\mapsto (x, y + \delta/2) \quad \text{by } \bar{r}$$
$$\mapsto (x, y + \delta) \quad \text{by } t_{(0,\delta/2)}.$$

That is,

$$t_{(0,\delta/2)}\bar{r}t_{(0,\delta/2)}^{-1} \cdot \bar{r}(x, y) = t_{(0,\delta)}(x, y).$$

By suitable choice of origin we can assume that the given rotation is r_θ. I claim that r_θ is the product of the reflection $\bar{r}$ in the x-axis and the reflection $r_{\theta/2}\bar{r}r_{\theta/2}^{-1}$ in the line obtained by rotating the x-axis through angle $\theta/2$. This also is intuitively obvious (Figure 1.3), and is confirmed by the following calculation:

$$(x, y) \mapsto (x, -y) \quad \text{by } \bar{r}$$
$$\mapsto (x \cos\theta/2 - y \sin\theta/2, -x \sin\theta/2 - y \cos\theta/2) \quad \text{by } r_{\theta/2}^{-1}$$
$$\mapsto (x \cos\theta/2 - y \sin\theta/2, x \sin\theta/2 + y \cos\theta/2) \quad \text{by } \bar{r}$$
$$\mapsto (x \cos^2\theta/2 - y \sin\theta/2 \cos\theta/2 - x \sin^2\theta/2 - y \sin\theta/2 \cos\theta/2,$$

$$x \sin \theta/2 \cos \theta/2 - y \sin^2 \theta/2 + x \sin \theta/2 \cos \theta/2$$
$$+ y \cos^2 \theta/2) \quad \text{by } r_{\theta/2}$$
$$= (x \cos \theta - y \sin \theta, x \sin \theta + y \cos \theta).$$

That is,

$$r_{\theta/2} \bar{r} r_{\theta/2}^{-1} \cdot \bar{r}(x, y) = r_\theta(x, y).$$

Conversely, suppose we have reflections $\bar{r}_L$, $\bar{r}_M$ in lines L, M. We choose L to be the x-axis, so $\bar{r}_L = \bar{r}$, and if L meets M, we choose their intersection to be O. In this case $\bar{r}_M \bar{r}_L = r_\theta$ by the second calculation above (taking $\theta/2$ to be the angle between L and M). If L does not meet M, then $\bar{r}_M \bar{r}_L = t_{(0,\delta)}$ by the first calculation above.   □

Corollary 1. *If $\bar{r}_M \bar{r}_L$ is a rotation, then $\bar{r}_{M'} \bar{r}_{L'} = \bar{r}_M \bar{r}_L$ for any lines L', M' with the same intersection as L, M and the same (signed) angle from L to M. If $\bar{r}_M \bar{r}_L$ is a translation, then $\bar{r}_{M'} \bar{r}_{L'} = \bar{r}_M \bar{r}_L$ for any lines L', M' parallel to L, M and the same (signed) distance apart.*

Proof. First take L, M to be the x-axis and the line through O at angle $\theta/2$ respectively, so

$$\bar{r}_M \bar{r}_L = r_{\theta/2} \bar{r} r_{\theta/2}^{-1} \cdot \bar{r} = r_\theta \quad \text{by the theorem.}$$

Take L', M' to be the lines through O at angles ϕ, $\theta/2 + \phi$, respectively, so $\bar{r}_{L'} = r_\phi \bar{r} r_\phi^{-1}$, $\bar{r}_{M'} = r_{\theta/2+\phi} \bar{r} r_{\theta/2+\phi}^{-1}$. Then

$$\bar{r}_{M'} \bar{r}_{L'} = r_{\theta/2+\phi} \bar{r} r_{\theta/2+\phi}^{-1} \cdot r_\phi \bar{r} r_\phi^{-1}$$
$$= r_\phi \cdot r_{\theta/2} \bar{r} r_{\theta/2}^{-1} \bar{r} \cdot r_\phi^{-1}$$
$$= r_\phi r_\theta r_\phi^{-1} = r_\theta = \bar{r}_M \bar{r}_L.$$

Second, take L, M to be the x-axis and the line $y = \delta/2$. Then if we take L', M' to be the lines $y = \gamma$, $y = \gamma + \delta/2$, a similar calculation shows $\bar{r}_{M'} \bar{r}_{L'} = \bar{r}_M \bar{r}_L$.   □

Corollary 2. *The set of translations and rotations is closed under product.*

Proof. The product of two translations is obviously a translation. To find the product of rotations $r_{P,\theta}$ (about P by θ) and $r_{Q,\phi}$ (about Q by ϕ), we express them as products of reflections in the lines L, M, N shown in Figure 1.4. (Such a choice of lines is always possible, by Corollary 1.) Then

$$r_{Q,\phi} \cdot r_{P,\theta} = \bar{r}_N \bar{r}_M \cdot \bar{r}_M \bar{r}_L = \bar{r}_N \bar{r}_L,$$

which is a rotation if N meets L and a translation otherwise.

The product of a translation and a rotation is similarly found to be a product of two reflections, by a suitable choice of lines.   □

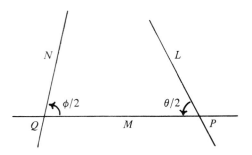

FIGURE 1.4.

Remarks. (1) A more concise way of computing with isometries is to express them as complex functions of one variable. We view the point $(x, y) \in \mathbb{R}^2$ as $z = x + iy \in \mathbb{C}$. Then $t_{(\alpha,\beta)}$ becomes the function $t_{\alpha+i\beta}(z) = \alpha + i\beta + z$, r_θ becomes the function $r_\theta(z) = e^{i\theta}z$, and $\bar{r}$ becomes the function $\bar{r}(z) = \bar{z}$. This greatly simplifies products of isometries (see Exercises) though it is only of peripheral interest in this chapter. The representation of isometries by complex functions is much more significant in spherical and hyperbolic geometry (see Chapters 3 and 4).

(2) Making a convenient choice of origin and axes (or equivalently, moving them by translation and rotation) makes many proofs easier by simplifying the coordinates. Another important example is the following.

Triangle inequality. *For any points P_1, P_2, P_3*

$$d(P_1, P_3) \leq d(P_1, P_2) + d(P_2, P_3).$$

Proof. By translating P_1 to O, then rotating about O until P_2 is on the x-axis, we can arrange that $P_1 = (0,0)$, $P_2 = (x_2, 0)$, $P_3 = (x_3, y_3)$. Hence, the inequality we have to prove becomes

$$\sqrt{x_3^2 + y_3^2} \leq x_2 + \sqrt{(x_3 - x_2)^2 + y_3^2}.$$

Squaring both sides and dividing by 2, we find the equivalent inequality

$$x_2 x_3 - x_2^2 \leq x_2\sqrt{(x_3 - x_2)^2 + y_3^2}.$$

Squaring again, and dividing by x_2^2 (if $x_2 = 0$, the inequality is already obvious), we get the equivalent inequality

$$(x_3 - x_2)^2 \leq (x_3 - x_2)^2 + y_3^2,$$

which is certainly true. $\square$

Exercises

1.3.1. Give reasons why the isometry $\bar{r}t_{(1,0)}$ is not the product of one or two reflections.

1.3.2. Check that reflection in the y-axis sends (x, y) to $(-x, y)$.

1.3.3. Check that $r_\theta = r_{\theta/2}\bar{r}r_{\theta/2}^{-1}\bar{r}$ using $r_\theta(z) = e^{i\theta}z$, $\bar{r}(z) = \bar{z}$.

1.3.4. Show that the complex form of a product of translations, rotations, and reflections is a function $f(z) = c + e^{i\theta}z$ or $\bar{r}(z) = c + e^{i\theta}\bar{z}$, where $c \in \mathbb{C}$, $\theta \in \mathbb{R}$. Show, conversely, that any function of this form is realizable as a product of isometries.

1.3.5. By suitable change of coordinate system, show that $f(z) = c + e^{i\theta}z$ represents a translation if $e^{i\theta} = 1$, and otherwise a rotation about the point $c/(1 - e^{i\theta})$.

1.3.6. Show that $t_c\bar{r} = \bar{r}t_c$, $r_\theta\bar{r} = \bar{r}r_{-\theta}$, $r_\theta t_c = t_{e^{i\theta}c}r_\theta$.

1.4   The Three Reflections Theorem

The reason for the fact, claimed in the previous section, that any isometry is the product of one, two, or three reflections is that any isometry is determined by its effect on a triangle. To be precise, we have the following:

Lemma. *Any isometry f of $\mathbb{R}^2$ is determined by the images $f(A)$, $f(B)$, $f(C)$ of three points A, B, C not in a line.*

Proof. It will suffice to show that any point $P \in \mathbb{R}^2$ is determined by its distances from A, B, C because $f(P)$ will then be determined by (the same) distances from $f(A)$, $f(B)$, $f(C)$. (The latter points are also not in a line because lines are preserved by isometries.)

This key fact is a consequence of the characterization of lines mentioned in Exercise 1.2.2: a line is the set of points equidistant from two given points. For completeness we now give a proof of this fact.

By suitable choice of axes we can assume that the given line L is $y = 0$. But then L is the set of points equidistant from $P = (0, -\alpha)$ and $Q = (0, \alpha)$ because

$$(x, y) \text{ is equidistant from } P \text{ and } Q$$
$$\Leftrightarrow x^2 + (y + \alpha)^2 = x^2 + (y - \alpha)^2$$
$$\Leftrightarrow 2\alpha y = -2\alpha y$$
$$\Leftrightarrow y = 0, \text{ since } \alpha \neq 0 \text{ for } P, Q \text{ to be distinct.}$$

Conversely, any two points P, Q can be assumed to be of the form $(0, -\alpha)$, $(0, \alpha)$ by choosing PQ to be the y-axis and O to be the midpoint of PQ.

The above calculation then shows that the set of points equidistant from P and Q is a line, namely, $y = 0$.

Now to prove the lemma, assume on the contrary that P, Q are two distinct points with the same distances from A, B, C. Then A, B, C are all in the set of points equidistant from P and Q; hence they lie in a line, contrary to the assumption. $\qquad\square$

Corollary. *If L is the line of points equidistant from P and Q, then reflection in L exchanges P and Q.*

Proof. As in the lemma we can assume $P = (0, -\alpha)$ and $Q = (0, \alpha)$, from which it follows that L is the line $y = 0$. But then reflection in L is $\bar{r}$, which sends (x, y) to $(x, -y)$, and hence exchanges P and Q. $\qquad\square$

Theorem. *Any isometry f of $\mathbb{R}^2$ is the product of one, two, or three reflections.*

Proof. Choose three points A, B, C not in a line and consider their f-images $f(A)$, $f(B)$, $f(C)$.

If two of A, B, C coincide with their f-images, say $A = f(A)$ and $B = f(B)$, then reflection in the line L through A, B must send C to $f(C)$ by the corollary because C and $f(C)$ are equidistant from $A = f(A)$ and equidistant from $B = f(B)$. This reflection, therefore, sends A, B, C to $f(A)$, $f(B)$, $f(C)$, respectively, and hence coincides with f by the lemma.

If one of A, B, C coincides with its f-image, say $A = f(A)$, first perform the reflection $\bar{g}$ in the line M of points equidistant from B and $f(B)$. Since $A = f(A)$ is equidistant from B and $f(B)$, hence on M, it is fixed by $\bar{g}$, so $\bar{g}$ sends A, B to $f(A)$, $f(B)$, respectively. If $\bar{g}$ also sends C to $f(C)$, we are finished. If not, we perform a second reflection $\bar{h}$ in the line through $f(A)$, $f(B)$ and conclude as above that $\bar{h}\bar{g}(C) = f(C)$. Then $\bar{h}\bar{g}$ sends A, B, C to $f(A)$, $f(B)$, $f(C)$, respectively, and hence coincides with f by the lemma.

Finally, if none of A, B, C coincides with its f-image, then we perform up to three reflections, in the lines L, M, N, respectively, equidistant from A and $f(A)$, B and $f(B)$, C and $f(C)$. One sees similarly that the product of one, two, or three of these reflections sends A to $f(A)$, B to $f(B)$, and C to $f(C)$, and hence coincides with f by the lemma. $\qquad\square$

Corollary. *The isometries of $\mathbb{R}^2$ form a group $\mathrm{Iso}(\mathbb{R}^2)$, and the products of even numbers of reflections form a subgroup $\mathrm{Iso}^+(\mathbb{R}^2)$ of index 2.*

Proof. It is clear that associativity holds for products of isometries (as for products of any maps) and that we have an identity isometry. The theorem establishes the less obvious condition—the existence of an inverse for each isometry. Indeed, since reflection $\bar{r}_L$ in any line L is self-inverse, the inverse of an isometry $\bar{r}_{L_1} \ldots \bar{r}_{L_n}$ is simply $\bar{r}_{L_n} \ldots \bar{r}_{L_1}$. Thus, the isometries form a group, $\mathrm{Iso}(\mathbb{R}^2)$.

It follows that the products $\bar{r}_{L_1} \ldots \bar{r}_{L_{2n}}$ of even numbers of reflections form a subgroup, $\mathrm{Iso}^+(\mathbb{R}^2)$ because products and inverses of such isometries are again of the same form.

To see that $\mathrm{Iso}^+(\mathbb{R}^2)$ is of index 2, we have to show that the coset $\mathrm{Iso}^+(\mathbb{R}^2) \cdot \bar{r} = \{$products of odd numbers of reflections$\}$ is not $\mathrm{Iso}^+(\mathbb{R}^2)$, i.e., that $\bar{r} \notin \mathrm{Iso}^+(\mathbb{R}^2)$. This follows from Corollary 2 to the theorem in Section 1.3, which says that $\mathrm{Iso}^+(\mathbb{R}^2)$ consists of rotations and translations. The fixed point set of a nontrivial rotation is a single point, the fixed point set of a nontrivial translation is empty, whereas the fixed point set of $\bar{r}$ is a line.   □

Remarks. (1) It is intuitively clear that the product of an even number of reflections preserves the sense of a clockwise oriented circle in $\mathbb{R}^2$, whereas the product of an odd number of reflections reverses it. The corollary shows that the classes $\mathrm{Iso}^+(\mathbb{R}^2)$ of even products and $\mathrm{Iso}^+(\mathbb{R}^2) \cdot \bar{r}$ of odd products are indeed mutually exclusive. Hence, it is meaningful to distinguish between *orientation-preserving isometries* (even products of reflections) and *orientation-reversing isometries* (odd products of reflections).

(2) In classical geometry, triangles Δ, Δ' were called *congruent* if their three side lengths were the same. Euclid assumed that in this case Δ could be "moved" to coincide with Δ'. The proof of the theorem shows, in fact, that the whole plane can be moved (by an isometry) so that Δ coincides with Δ'. Hence, it is now appropriate to define any figures Φ, Φ' to be *congruent* if there is an isometry mapping Φ onto Φ'.

Exercises

1.4.1. Show that any isometry of $\mathbb{R}^3$ is the product of one, two, three, or four reflections.

1.4.2. Show that the translations form a group but that the rotations do not.

1.4.3. If $f_1 = t_{c_1} r_{\theta_1}$ and $f_2 = t_{c_2} r_{\theta_2}$ with $c_1, c_2 \in \mathbb{C}$ and $\theta_1, \theta_2 \in \mathbb{R}$, show that

$$f_1 f_2 = f_2 f_1 \Leftrightarrow c_2(1 - e^{i\theta_1}) = c_1(1 - e^{i\theta_2}).$$

Deduce that two orientation-preserving isometries commute if and only if they are both translations, or both rotations about the same point.

1.5   Orientation-Reversing Isometries

The three reflections theorem and its corollary tell us that the orientation-preserving isometries of $\mathbb{R}^2$ are the products of two reflections, which we know from the theorem in Section 1.3 to be the rotations and translations.

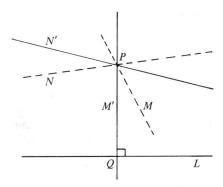

FIGURE 1.5.

They also tell us that the orientation-reversing isometries are products of one or three reflections. Thus, to complete our understanding of euclidean isometries we seek a more illuminating description of products of three reflections.

Each of these turns out to be of the same geometric type—the product of a reflection with a translation in the direction of the line of reflection L—called a *glide reflection* with *axis* L. An ordinary reflection is thus the special case of a glide reflection with trivial translation. The general glide reflection $\bar{f}$ takes its simplest form when the axis is the x-axis, in which case $\bar{f} = t_{(\alpha,0)}\bar{r}$, or in terms of the complex coordinate, $\bar{f}(z) = \alpha + \bar{z}$.

Theorem. *A product $\bar{r}_N \bar{r}_M \bar{r}_L$ of reflections in lines L, M, N is a glide reflection.*

Proof. Case (i). L, M, N have a common point or are parallel.

First suppose L, M, N have a common point P. Then (by Section 1.3) $\bar{r}_M \bar{r}_L$ is a rotation about P, and $\bar{r}_M \bar{r}_L = \bar{r}_{M'} \bar{r}_{L'}$ for any other lines L', M' through P separated by the same (signed) angle. In particular we can choose $M' = N$ which gives

$$\bar{r}_N \bar{r}_M \bar{r}_L = \bar{r}_N \bar{r}_N \bar{r}_{L'} = \bar{r}_{L'},$$

so $\bar{r}_N \bar{r}_M \bar{r}_L$ is a reflection, hence trivially a glide reflection. Now suppose L, M, N are parallel. Then (by Section 1.3 again) $\bar{r}_M \bar{r}_L = \bar{r}_{M'} \bar{r}_{L'}$ for any lines L', M' parallel to L, M and separated by the same (signed) distance. In particular, we can choose $M' = N$, and the same calculation shows that $\bar{r}_N \bar{r}_M \bar{r}_L$ is a glide reflection.

Case (ii). First assume that the three lines are such that L does not pass through the intersection P of M and N.

Then $\bar{r}_N \bar{r}_M$ is a rotation about P and $\bar{r}_N \bar{r}_M = \bar{r}_{N'} \bar{r}_{M'}$ for any other lines M', N' through P separated by the same angle. We choose M' per-

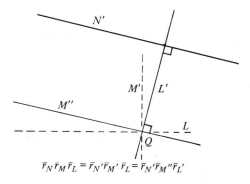

$$\bar{r}_N\bar{r}_M\bar{r}_L = \bar{r}_{N'}\bar{r}_{M'}\bar{r}_L = \bar{r}_{N'}\bar{r}_{M''}\bar{r}_{L'}$$

FIGURE 1.6.

pendicular to L, at Q say (Figure 1.5), so

$$\bar{r}_N\bar{r}_M\bar{r}_L = \bar{r}_{N'}\bar{r}_{M'}\bar{r}_L, \quad \text{where } M', L \text{ are perpendicular.}$$

Now $\bar{r}_{M'}\bar{r}_L$ is a rotation about Q and $\bar{r}_{M'}\bar{r}_L = \bar{r}_{M''}\bar{r}_{L'}$ for any (appropriately oriented) perpendicular lines M'', L' through Q. In particular, we can choose L' perpendicular to N', at R say (Figure 1.6). Then

$$\bar{r}_N\bar{r}_M\bar{r}_L = \bar{r}_{N'}\bar{r}_{M'}\bar{r}_L = \bar{r}_{N'}\bar{r}_{M''}\bar{r}_{L'}$$

and, since L' is the common perpendicular of M'' and N', $\bar{r}_{N'}\bar{r}_{M''}$ is a translation in the direction of L'. Consequently, $\bar{r}_{N'}\bar{r}_{M''} \cdot \bar{r}_{L'}$ is a glide reflection with axis L'.

The argument is similar if M or N is the line that does not pass through the intersection of the other two. Rotate pairs of intersecting lines about their common point until the first (or last) line is a common perpendicular of the other two.   □

With this theorem, together with those of Sections 1.3 and 1.4, we have obtained the following result:

Classification of Euclidean Isometries. *Each isometry of $\mathbb{R}^2$ is either a rotation, translation, or glide reflection.*

Remark. A somewhat simpler, analytic proof that orientation-reversing isometries are glide reflections can be obtained from their representation as complex functions (see Exercise 1.5.1). We have opted for the synthetic proof above because it also works, with appropriate modifications, in spherical and hyperbolic geometry (see Exercise 3.6.1 and Section 4.5).

Exercises

1.5.1. Show that each orientation-reversing isometry has the form

$$\bar{f}(z) = c + e^{i\theta}\bar{z}, \quad \text{which becomes}$$
$$\bar{g}(z) = ce^{-i\theta/2} + \bar{z} \quad \text{after rotation of axes and}$$
$$\bar{h}(z) = = \alpha + \bar{z}, \quad \text{where } \alpha \in \mathbb{R}, \text{ after translation of } O.$$

1.5.2. Find the axis of $\bar{f}(z) = c + e^{i\theta}\bar{z}$, and the length α of translation.

1.5.3. Deduce from the classification that each euclidean isometry has exactly one of the following:

(i) A line of fixed points.

(ii) A single fixed point.

(iii) No fixed points, and a parallel family of invariant lines (an invariant line is a line mapped onto itself by the isometry).

(iv) No fixed points, and a single invariant line.

1.5.4. Show that an isometry g has the property listed in Exercise 1.5.3 respectively when

(i) $g = \bar{r}_L$.

(ii) $g = \bar{r}_M\bar{r}_L$, where L, M intersect.

(iii) $g = \bar{r}_M\bar{r}_L$, where L, M are parallel.

(iv) $g = \bar{r}_N\bar{r}_M\bar{r}_L$, where L, M, N have no common point and are not parallel.

1.6   Distinctive Features of Euclidean Geometry

The three basic two-dimensional geometries—euclidean, spherical, and hyperbolic—all have isometries that can be regarded as reflections in lines. Of course, the notions of "isometry" and "line" depend on the distance function for the geometry in question, but if we define a reflection to be an isometry whose fixed point set is a line, then reflections occur in all three geometries. Moreover, we shall prove (as we already have done in the euclidean case) that they generate all the isometries. In particular, each geometry has "rotations", which are products of reflections in two intersecting lines.

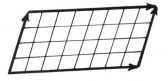

FIGURE 1.7.

Such are the common features of the three geometries. However, the differences are more interesting. Since all isometries are generated by reflections in lines, geometric differences emerge from differences in the properties of lines. These can be described in terms of departures from properties of euclidean lines, the most distinctive of which are the following. (The properties claimed for the euclidean plane are well-known or easy to prove. The properties of the sphere and hyperbolic plane will be proved in Chapters 3 and 4.)

1.6.1   Existence of Parallels

In the euclidean plane, for each line L and point $P \notin L$ there is a unique line L' through P which does not meet L. L' is called the *parallel* to L through P. Parallels provide us with a global notion of *direction* in the euclidean plane. Each member of a family of parallels has the same direction, measured by the angle a member of the family makes with the x-axis, and parallels are a constant distance apart.

A euclidean translation slides each member of a family of parallels along itself a constant distance. Consequently, translations always commute (Figure 1.7).

In the sphere, "lines" are great circles, and hence any two of them intersect. Thus, there are no parallels, no global notion of direction (which way is north at the north pole?), and no translations. The only orientation-preserving isometries are rotations. Each rotation slides just *one* line (great circle) along itself, together with the curves at constant distance from this line. These "equidistant curves", however, are not lines.

In the hyperbolic plane there are many lines L' through a point $P \notin L$ which do not meet L. (This is typical of the way hyperbolic geometry departs from euclidean—in the opposite way from spherical geometry.) Translations exist, but each translation slides just one line along itself, together with the curves at constant distance from this line. These "equidistant curves" are also not lines, and translations with different invariant lines do not commute.

1.6.2   The Angle Sum of a Triangle

In axiomatic geometry, all distinctively euclidean properties are consequences of the existence of parallels. One particularly elegant consequence is that the angle sum of a triangle is π. Figure 1.8 summarises the proof. The two angles marked α are equal because a translation carries one to the other. The two angles marked β at B are equal by rotation about B, and the external β is equal to the β at C by another translation. Looking now at C, we obviously have $\alpha + \beta + \gamma = \pi$.

In contrast, the angle sum of a spherical triangle is $> \pi$, and the angle sum of a hyperbolic triangle is $< \pi$.

By forming unions of euclidean triangles, one finds that the angle sum of a euclidean n-gon is $(n-2)\pi$ (see exercises). In particular, each angle of a *regular* n-gon is $\frac{n-2}{n}\pi$. Thus, the regular 3-gon (equilateral triangle) has angle $\pi/3$, the regular 4-gon (square) has angle $\pi/2$, the regular 6-gon (hexagon) has angle $2\pi/3$, and these are the only regular n-gons with angles of the form $2\pi/m$ for $m \in \mathbb{Z}$. It follows that the only tessellations of the euclidean plane by regular n-gons are the well-known ones by equilateral triangles, squares, and regular hexagons (Figure 1.9), since in any such tessellation the angle 2π at each vertex must be divided among an integral number of equal corner angles.

The tessellations of the sphere by regular spherical n-gons are also well-known—they correspond to the five regular polyhedra—but in the hyperbolic plane there are infinitely many possibilities.

It is also interesting to consider tessellations in which the basic tile is an *irregular* polygon, say a triangle with angles π/p, π/q, π/r. It turns out that such a triangle tessellates the

$$\text{euclidean plane} \Leftrightarrow \frac{1}{p} + \frac{1}{r} + \frac{1}{r} = 1,$$

$$\text{sphere} \Leftrightarrow \frac{1}{p} + \frac{1}{q} + \frac{1}{r} > 1,$$

$$\text{hyperbolic plane} \Leftrightarrow \frac{1}{p} + \frac{1}{q} + \frac{1}{r} < 1.$$

(See Chapters 7 and 8.)

1.6.3   Existence of Similarities

The euclidean plane admits mappings, called *similarities* or *dilatations*, which multiply all distances by a constant factor $\lambda \neq 0$. The typical similarity is $(x, y) \mapsto (\lambda x, \lambda y)$. Figures related by a similarity are said to be "of the same shape" or "similar." In particular, all triangles with the same angles are similar, as are all squares. The existence of squares of different sizes means, e.g., that n^2 unit squares fill a square of side n. This leads to

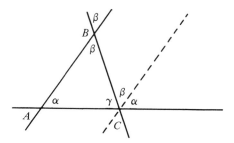

FIGURE 1.8.

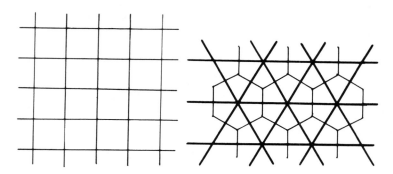

FIGURE 1.9.

the property of euclidean area that multiplying the lengths in a figure by λ multiplies its area by λ^2.

The euclidean plane is unique in having this simple dependence of area on length because the sphere and the hyperbolic plane do not admit similarities (except with $\lambda = 1$). There the relationships between length and area are more complicated—involving circular and hyperbolic functions, respectively—but the relationship between *angle* and area is delightfully simple.

This is a benefit of having the triangle's angle sum unequal to π. One then has a nontrivial angular excess function for triangles Δ:

$$\text{excess}(\Delta) = \text{angle sum}(\Delta) - \pi,$$

which is proportional to area because it is *additive*. That is, if triangle Δ is split into triangles Δ_1 and Δ_2, then

$$\text{excess}(\Delta) = \text{excess}(\Delta_1) + \text{excess}(\Delta_2).$$

This is easily verified by reference to Figure 1.10, using the fact that $\gamma_1 + \gamma_2 = \pi$. Euclidean geometry misses out on this property because it is too simple—its angular excess function is zero. It can be shown that any

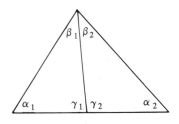

FIGURE 1.10.

continuous, nonzero, additive function must be proportional to area, and we shall later confirm directly that angular excess measures area in the sphere and hyperbolic plane.

The identification of angular excess with area clearly shows why similar triangles of different sizes cannot exist in the sphere and hyperbolic plane. Triangles with the same angles have the same area, and hence one cannot be larger than the other.

Exercises

1.6.1. Show, by prolonging the sides of an arbitrary polygon Π, that Π can be divided into convex polygons. Hence, show that Π can be divided into a finite number of triangles.

1.6.2. Deduce from 1.6.1 that the angle sum of a euclidean n-gon Π is $(n-2)\pi$.

1.6.3. Deduce also that for a spherical or hyperbolic polygon Π the angular excess,

$$\text{excess}(\Pi) = \text{angle sum}(\Pi) - (n-2)\pi,$$

is additive.

1.6.4. Show that the mappings of $\mathbb{C}$ generated by orientation-preserving euclidean isometries and similarities are the functions of the form $f(z) = az + b$.

1.7   Discussion

The idea of doing geometry in terms of numbers and equations caught on after the publication of Descartes' *La Géométrie* [1637]. However, as mentioned above, the idea that numbers and equations *are* geometric objects arose much later. In fact, the idea had no solid foundation until 1858, when the set $\mathbb{R}$ of real numbers was first given a clear definition, by Dedekind (published in his [1872]). Dedekind's definition explains in particular the *continuity* of $\mathbb{R}$ which enables it to serve as a model for the line.

Once one has this model for the line it is relatively straightforward to model the plane by $\mathbb{R}^2$ and to verify Euclid's axioms. This was first done in detail by Hilbert [1899], thus subordinating geometry to the number concept after 2000 years of independence. It should be mentioned, however, that any construction of $\mathbb{R}$ from the natural numbers $0, 1, 2, \ldots$ involves infinite sets. For example, a Dedekind real number is a partition of the rational numbers into two sets L, U where each member of L is less than every member of U. (Other definitions, by Méray, Cantor, and Weierstrass, are even more complicated than this, but they give isomorphic sets $\mathbb{R}$.) Thus, a "point" is a much more subtle object than naive intuition suggests. One is reminded of the situation in quantum physics where even empty space turns out to be complicated!

Be that as it may, the idea of interpreting points as numbers has been unexpectedly fruitful. Of course, we expect $\mathbb{R}$ to behave like a line because $\mathbb{R}$ was constructed with that purpose in mind, and it is no surprise that $+$ and $\times$ have a geometric meaning on the line ($+$ as translation, $\times$ as dilatation). It may also not be a surprise that $+$ has a meaning in $\mathbb{R}^2$ (translation $=$ vector addition) and so does multiplication by a real number (dilatation again). But it is surely an unexpected bonus when multiplication by a *complex* number turns out to be geometrically meaningful.

After all, this multiplication is forced on us by algebra—by the demand that $i^2 = -1$ and that the field laws hold—yet when $a + ib \in \mathbb{C}$ is interpreted as $(a, b) \in \mathbb{R}^2$, multiplication by a complex number is simply the product of a dilatation and a rotation. In particular, we have the miraculous fact that multiplication by $e^{i\theta}$ is rotation through θ. And this is just the beginning of the interplay between complex numbers and angles, leading to many applications of complex numbers, and particularly complex *functions*, in geometry.

As Riemann [1851] observed, the fundamental property of a differentiable complex function f is that f preserves angles. If $[f(z_2) - f(z_1)]/(z_2 - z_1)$ has the same limit no matter how z_2 tends to z_1, then the distances $|f(z_2) - f(z_1)|$ and $|z_2 - z_1|$ must have the same limiting ratio. In a neighborhood where this limiting ratio is nonzero, f is, therefore, a "similarity in the limit" and, in particular, f preserves angles because angles are determined by the limiting ratios of sides of triangles. Differentiability is, therefore, a much stronger constraint on complex functions than it is on real functions, and the study of differentiable complex functions is in principle a part of geometry. It is the study of *conformal* mappings—mappings which preserve the magnitude and sign of angles. (We should also mention the converse theorem of Riemann that conformal mappings are given by differentiable complex functions. Of course, in geometry one normally complements the conformal mappings by the *anticonformal* mappings—those which preserve the magnitude of angles but reverse their sign. These correspond to differentiable functions of $\bar{z}$.)

This claim is borne out by theorems of complex analysis which show that the differentiable functions on particular complex domains comprise very restricted, geometrically meaningful classes. A number of such theorems (which go back to Riemann [1851]) may be found in Jones and Singerman [1987, p. 200]. The one relevant to this chapter is the following. *The differentiable functions mapping $\mathbb{C}$ onto $\mathbb{C}$ are precisely the functions* $f(z) = az + b$. Recall from Exercise 1.6.4 that these are just the products of dilatations with orientation-preserving euclidean isometries. Thus, the theorem can be restated geometrically as follows: *any conformal map of the plane onto itself is the product of a dilatation with an orientation-preserving euclidean isometry.* Thus, if we relax the definition of euclidean mapping to include dilatations—a reasonable step because euclidean geometry is really concerned with lengths as ratios rather than as absolutes—then we have a simple new characterization of euclidean plane geometry. It is the geometry of angle-preserving maps of $\mathbb{C}$ onto itself.

2

Euclidean Surfaces

2.1 Euclid on Manifolds

The aim of this chapter is to answer the question: which unbounded surfaces look locally like the euclidean plane $\mathbb{R}^2$? The question arises because $\mathbb{R}^2$ is intended to model "flat" surfaces in the real world; yet all physical flat surfaces are of finite extent and have boundaries. It is not clear that such a surface would resemble $\mathbb{R}^2$ when extended indefinitely, even if small parts of it matched small parts of $\mathbb{R}^2$ with absolute precision. Indeed, we may never know enough about the large-scale structure of the universe to say what an unbounded flat surface would really be like. What we can do, however, is find which flat surfaces are *mathematically* possible.

In doing so, we shall meet, in a simple but nontrivial form, one of the most fruitful ideas of modern mathematics—the concept of *manifold*. In brief, an n-dimensional manifold is a space S in which each point has a neighborhood "like" an open ball in the euclidean space $\mathbb{R}^n$. This allows us to study complicated spaces in terms of familiar properties of $\mathbb{R}^n$. Which properties of $\mathbb{R}^n$ we can use depends on how "alike" the neighborhoods in S and $\mathbb{R}^n$ are. At one extreme, they may be merely homeomorphic, in which case S is a *topological* manifold, and we can only use topological properties of $\mathbb{R}^n$. We are going to consider the other extreme, where the neighborhoods are isometric. In this case, S is a *euclidean* manifold, and all concepts which have meaning in neighborhoods of $\mathbb{R}^n$, such as length, angle, and straightness, also apply to S.

In the terminology of manifolds, the question with which we began this section concerns the two-dimensional euclidean manifolds. Actually, we shall impose additional, geometrically natural conditions on our manifolds before we finally answer this question (see Section 2.7). And before that we will look at some examples to get an idea of the direction in which the answer should be sought.

2.2 The Cylinder

It is intuitively clear that the cylinder is "locally like" the plane $\mathbb{R}^2$ because a cylinder can be made by joining the edges of a strip of paper, and a strip of paper is like part of the plane. How do we formalize this idea?

The most brutally direct way is to take a strip S of $\mathbb{R}^2$ bounded by parallel lines, say $x = 0$ and $x = 1$, and say that the points of the cylinder C *are* the points of S, with the additional proviso that points $(0, y)$ and $(1, y)$ on S are the *same* point on C (this "joins" the two edges of the strip, cf. Figure 2.1). However, this construction is inelegant because the "join" shows—it consists of the points on the cylinder which are pairs of points on the strip. We therefore elaborate the construction so as to treat all points of the cylinder equally, by using all points of $\mathbb{R}^2$.

Along with each point (x, y) in the strip we take all points $(x + n, y)$, at integer horizontal distances from it, to represent the same point of C (one such set of points is indicated by stars in Figure 2.2). Intuitively speaking, this is a "seamless" construction because we form C by "rolling up" the whole plane.

The process can be described abstractly as construction of the *quotient space* or *orbit space* $\mathbb{R}^2/\Gamma$, where Γ is the group of integer horizontal translations of $\mathbb{R}^2$. A *point* of $C = \mathbb{R}^2/\Gamma$ is a set of the form $\{(x + n, y) \mid n \in \mathbb{Z}\}$, called the Γ-*orbit* of (x, y), hence the term "orbit space." It is natural to use the abbreviation ΓP for the Γ-orbit of the point P because

$$\Gamma P = \{gP \mid g \in \Gamma\}.$$

[We are also abbreviating $g(P)$ by gP.]

A convenient way to visualize the orbit space $\mathbb{R}^2/\Gamma$ is to focus on a *fundamental region*—a part of the plane which contains a representative of each Γ-orbit with at most one representative of each Γ-orbit in its interior. One fundamental region is the strip S we considered first. We can now see another reason why S is inelegant—it is quite an arbitrary choice among infinitely many possible fundamental regions. But at least S enables us to visualize the quotient $\mathbb{R}^2/\Gamma$, namely, as the result of joining those points on the boundary of S which belong to the same Γ-orbit.

To complete the description of $C = \mathbb{R}^2/\Gamma$ we have to give a meaning to "distance" in C, and, of course, we intend this to be the same as distance

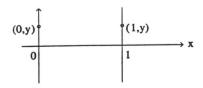

FIGURE 2.1.

FIGURE 2.2.

in $\mathbb{R}^2$ for "sufficiently small neighborhoods" in C. We let

$$d_C(\Gamma P, \Gamma Q) = \min\{d(P', Q') \mid P' \in \Gamma P,\ Q' \in \Gamma Q\},$$

where d denotes the (euclidean) distance in $\mathbb{R}^2$. The right-hand side also equals

$$\min\{d(P, Q') \mid Q' \in \Gamma Q\}$$

because each $P' \in \Gamma P$ has the same set of distances to the members of ΓQ. The latter expression shows that d_C is well-defined because for each $P \in \mathbb{R}^2$ there is a nearest $Q' \in \Gamma Q$ (possibly one of a pair that are equally near). Notice that if $d(P, Q) < 1/2$ we have $d_C(\Gamma P, \Gamma Q) = d(P, Q)$ because in this case Q is nearest to P among the members of ΓQ.

The map which sends each $P \in \mathbb{R}^2$ to its Γ-orbit $\Gamma P \in \mathbb{R}^2/\Gamma$ is called the *orbit map*, and we denote it by $\Gamma\cdot$ because it is formally multiplication on the left by Γ. By the observation just made, $\Gamma\cdot$ maps any disc $D \subset \mathbb{R}^2$ of diameter $< 1/2$ isometrically into $\mathbb{R}^2/\Gamma$; thus, $\Gamma\cdot$ is a *local isometry*. Figure 2.3 gives an intuitive picture of the orbit map by representing an orbit (set of stars in $\mathbb{R}^2$) belonging to the abstract cylinder $\mathbb{R}^2/\Gamma$ by a point (single star) belonging to a concrete cylinder in space. Γ sends all the stars in $\mathbb{R}^2$ to the single star on the cylinder and, therefore, "wraps the plane around the cylinder". For this reason, $\Gamma\cdot$ is called a *covering* of $\mathbb{R}^2/\Gamma$ by $\mathbb{R}^2$.

The local isometry property of the orbit map means that, within discs of diameter $< 1/2$, the geometry of the cylinder is the same as the geometry of the plane. However, interesting differences emerge when we try to extend geometric concepts to the whole cylinder. For example, it is natural to define a *line* on C to be the $\Gamma\cdot$-image of a line on $\mathbb{R}^2$. Such "lines" are locally the same as ordinary lines—their intersections with discs of diameter $< 1/2$ are line segments—but they can be globally quite different. There are three distinct types of line on the cylinder, illustrated in Figure 2.4, and consequently a variety of ways in which two or more lines can interact (see Exercises).

The cylinder C is usually conceived as a surface lying in three-dimensional space, as we have done in Figures 2.3 and 2.4. This view helps us to anticipate some of the cylinder's geometric properties, which can then be checked against the formal definition of C as the quotient $\mathbb{R}^2/\Gamma$. The definition of C as $\mathbb{R}^2/\Gamma$ has the advantage of giving the most direct definition of distance

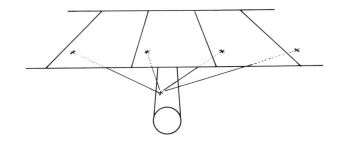

FIGURE 2.3.

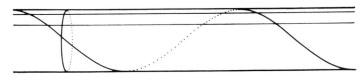

FIGURE 2.4.

on C and is itself quite helpful to intuition, corresponding as it does to the view of C as a repeating planar pattern (Figure 2.2), a *tessellation* of $\mathbb{R}^2$ by the images of a fundamental region for Γ.

The really decisive advantage of the quotient construction comes with the examples of euclidean surfaces which follow in Sections 2.3 and 2.4, as they *cannot* be represented properly in three-dimensional space.

Exercises

2.2.1. Which of the following properties of euclidean lines hold for lines on the cylinder?

(i) There is a line through any two points.

(ii) There is a unique line through any two points.

(iii) Two lines meet in at most one point.

(iv) There are lines which do not meet.

(v) A line has infinite length.

(vi) A line gives the shortest distance between two points.

(vii) A line does not cross itself.

2.2.2. If t is a nontrivial translation of $\mathbb{R}^2$, and Γ is the group $\langle t \rangle$ generated by t, define $\mathbb{R}^2/\Gamma$ and show that it is the same as the cylinder above, up to a change of scale.

2.2.3. Show that a disc D of diameter $3/4$ is mapped one-to-one into C by the orbit map, but that $d_C(\Gamma P, \Gamma Q) \neq d(P, Q)$ for certain $P, Q \in D$.

2.3   The Twisted Cylinder

The twisted cylinder C^* is constructed by joining opposite sides of a parallel-sided strip S, but with a twist. The resulting surface cannot lie in ordinary three-dimensional space without intersecting itself, though a fairly representative part of it can. This part is the Möbius band M, obtained by joining opposite sides of a rectangle R with a (half) twist (Figure 2.5). The twisted cylinder is obtained by prolonging the transverse line segments of M, just as the strip S is obtained by prolonging the transverse line segments of R.

This intuitive picture gives a glimpse of some of the curious properties of the twisted cylinder, which the reader may already know from the Möbius band (see Exercises). However, even more so than in the case of the ordinary cylinder, it is formally easier to define the twisted cylinder as a quotient $\mathbb{R}^2/\Gamma$.

The strip S is now the fundamental region of a group Γ generated by a glide reflection, which reflects in the x-axis and translates the x-axis by distance 1 (thus "joining" opposite sides of S with a twist). Figure 2.6 shows the images of S and, marked by stars, the Γ-orbit $\{(x+n, (-1)^n y) \mid n \in \mathbb{Z}\}$ of a typical point (x, y). As in Section 2.2 we define the points of $\mathbb{R}^2/\Gamma$ to be the Γ-orbits of points in $\mathbb{R}^2$, and use $\min\{d(P', Q') \mid P' \in \Gamma P, Q' \in \Gamma Q\}$ as the distance between any two points $\Gamma P, \Gamma Q \in C^*$. This distance is well defined for reasons similar to those that apply to the cylinder.

This makes C^* a euclidean surface, identical with $\mathbb{R}^2$ within discs of diameter less than $1/2$, but remarkably different from it in the large. For example, if we define *lines* in C^* to be the Γ-images of lines in $\mathbb{R}^2$, then the lines L, L' in Figure 2.7 have the same image line ΓL in C^*. This is

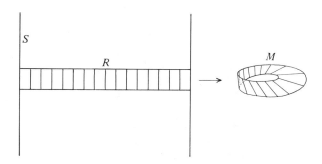

FIGURE 2.5.

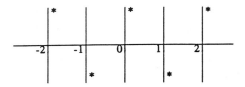

FIGURE 2.6.

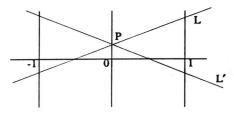

FIGURE 2.7.

because L' results from L by the glide reflection which generates Γ. But L and L' clearly cross at the point P; hence ΓL crosses *itself* at the Γ-image ΓP of P.

Exercises

2.3.1. Make a paper model of a Möbius band on which there is a line which crosses itself.

2.3.2. Show that there is a closed line L_0 of minimum length on C^* and that all other closed lines have twice the length of L_0.

2.3.3. Classify the isometries of the twisted cylinder. (Hint: An isometry of $C^* = \mathbb{R}^2/\Gamma$ is induced by an isometry of $\mathbb{R}^2$ which sends Γ-orbits to Γ-orbits. Each such isometry must map the x-axis onto itself.)

2.4   The Torus and the Klein Bottle

The torus is usually viewed as the doughnut-shaped surface obtained by rotating a circle in space. Such a surface can also be obtained by joining opposite sides of a rectangle (Figure 2.8). However, it is evident that distances on the rectangle are distorted by this construction—and we rather wish they were not—so again it is better to define the torus as a quotient $\mathbb{R}^2/\Gamma$ to carry over the local geometry of $\mathbb{R}^2$.

The most general Γ we can take has an arbitrary parallelogram as fundamental region (Figure 2.9). The generators of Γ are translations t_1, t_2 which carry the left side to the right, and the lower side to the upper, respectively.

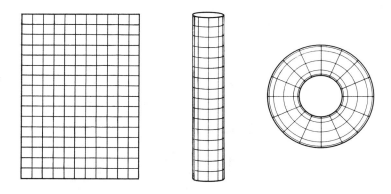

FIGURE 2.8.

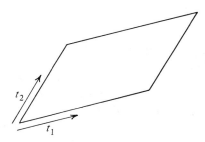

FIGURE 2.9.

Thus, $\mathbb{R}^2/\Gamma$ is the result of joining opposite sides of the parallelogram, and hence topologically the same as our "doughnut" torus constructed from a rectangle. (We do, however, get geometrically different tori from differently shaped parallelograms. An example of a shape-dependent property of the torus is given in Exercise 2.4.3.)

The Klein bottle is related to the torus in much the same way that the twisted cylinder is related to the cylinder. The usual construction is by joining opposite sides of a rectangle, with one pair of sides being joined with a twist. Figure 2.10 shows what happens to the rectangle whose sides have been labeled and directed so that the successive steps can be followed more easily. Like-labeled sides have to be joined, with their arrows pointing in the same direction.

Again it is clear that this construction distorts lengths, and it is also unfortunate that the surface intersects itself. As before, we resolve these difficulties by defining the Klein bottle as a quotient $\mathbb{R}^2/\Gamma$. The rectangle is the fundamental region for a group Γ generated by a glide reflection g in the horizontal direction and a translation t in the vertical direction (Figure 2.11). We need a rectangle in this case because the reflection axis of g has

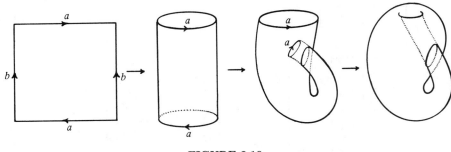

FIGURE 2.10.

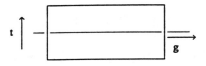

FIGURE 2.11.

to be perpendicular to the sides which g "joins with a twist," and hence perpendicular to the direction of t. This means that there is not such a variety of geometrically different Klein bottles as there are tori, though there are still infinitely many (one for each height-to-width ratio).

There is an apparent asymmetry in the definition of the Klein bottle because one of the generators given for the Klein bottle group Γ is a glide reflection and the other is not. However, one can equally well generate Γ by two glide reflections, for example, g and gt. The latter is a glide reflection, e.g., by the classification of isometries in Section 1.5.

As with the previous surfaces $\mathbb{R}^2/\Gamma$, we have an orbit map $\Gamma\cdot : \mathbb{R}^2 \to \mathbb{R}^2/\Gamma$ defined by $P \mapsto \Gamma P$ and a distance function on $\mathbb{R}^2/\Gamma$ which gives each orbit $P \in \mathbb{R}^2/\Gamma$ a neighborhood isometric to a euclidean disc.

Exercises

2.4.1. Show that a torus T has arbitrarily long closed lines and also infinitely long "open" lines.

2.4.2. Show that any translation of $\mathbb{R}^2$ induces an isometry of the torus $T = \mathbb{R}^2/\Gamma$.

2.4.3.

(i) If the fundamental region for T is a square, show that T has an isometry of order 4 induced by a rotation of $\mathbb{R}^2$.

(ii) If the fundamental region for T is a rhombus with angles $\pi/3$, $2\pi/3$, show that T has an isometry of order 6 induced by a rotation of $\mathbb{R}^2$.

2.4.4. Show that a Klein bottle can be cut into two Möbius bands.

2.4.5. Show that any torus or Klein bottle may be represented as a quotient of a cylinder.

2.5 Quotient Surfaces

The surfaces $S = \mathbb{R}^2/\Gamma$ constructed in Sections 2.2 to 2.4 all have a "locally euclidean structure" inherited from $\mathbb{R}^2$ via the Γ-orbit map. To be precise, there is a well-defined distance $d_S(\Gamma P, \Gamma Q)$ between orbits such that, for sufficiently small ϵ, the ϵ-neighbourhood of ΓP,

$$D_\epsilon(\Gamma P) = \{\Gamma Q \mid d_S(\Gamma P, \Gamma Q) < \epsilon\}$$

is isometric to $D_\epsilon(P)$ under the Γ-orbit map. A general definition of "euclidean surfaces" which encompasses the surfaces $\mathbb{R}^2/\Gamma$ of Sections 2.2 to 2.4 will be given in Section 2.7. Right now we wish to show that there are no more euclidean surfaces $\mathbb{R}^2/\Gamma$ by showing that any other group Γ yields a quotient which is not a surface.

Of course we can define $\mathbb{R}^2/\Gamma$ as *a set* (of Γ-orbits) for any group Γ of isometries of $\mathbb{R}^2$. But $\mathbb{R}^2/\Gamma$ is locally isometric to $\mathbb{R}^2$, under the orbit map, only for certain Γ. There is no meaningful distance at all between two different Γ-orbits which converge to the same point (a Γ with this property is described in Section 2.6). Even when $\mathbb{R}^2/\Gamma$ has a distance function, it may not behave everywhere like euclidean distance. For example, if Γ is the group generated by the rotation $r_{\pi/2}$ about O, then each neighborhood of O is mapped "4-to-1" into $\mathbb{R}^2/\Gamma$ in the manner indicated by Figure 2.12. This results, e.g., in the "circle" of radius ϵ centered on $\{0\} \in \mathbb{R}^2/\Gamma$ having circumference $\epsilon\pi/2$ instead of the euclidean value $2\pi\epsilon$.

To describe the groups Γ for which $\mathbb{R}^2/\Gamma$ is locally euclidean we make the following definitions.

We call Γ *discontinuous* if no $P \in \mathbb{R}^2$ has a Γ-orbit with a limit point (i.e., a point whose neighborhoods all include infinitely many points of ΓP), and *fixed point free* if $gP \neq P$ for each $P \in \mathbb{R}^2$ and each $g \neq 1$ (the identity element) in Γ.

Lemma. *If Γ is a group of isometries of $\mathbb{R}^2$, then Γ is discontinuous and fixed point free if and only if each $P \in \mathbb{R}^2$ has a neighborhood D_P in which each point belongs to a different Γ-orbit.*

Proof. Suppose Γ is discontinuous and fixed point free, and consider any $P \in \mathbb{R}^2$. Since Γ is discontinuous, there is a $\delta > 0$ such that all points in the Γ-orbit of P are at a distance $\geq \delta$ from P. Then, since Γ is fixed point free, gP is at a distance $\geq \delta$ from P for each $g \neq 1$ in Γ. Thus, the whole neighborhood D_P of P with radius $\delta/3$ is shifted to a position disjoint from

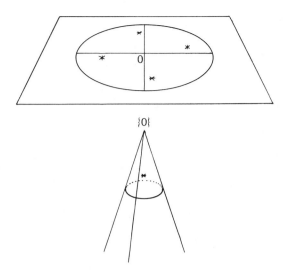

FIGURE 2.12.

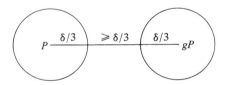

FIGURE 2.13.

D_P by g (Figure 2.13), hence D_P cannot contain two points in the same Γ-orbit.

Conversely, suppose each $P \in \mathbb{R}^2$ has a neighborhood D_P in which each point belongs to a different Γ-orbit. Then Γ must be discontinuous, otherwise some $P \in \mathbb{R}^2$ would have members of the same Γ-orbit in all its neighborhoods.

Also, Γ must be fixed point free. If not, consider a fixed point Q of some $g \neq 1$ in Γ. Since $g \neq 1$, g cannot be the identity on any neighborhood of Q (otherwise it would fix three points not in a line and hence would be the identity by the lemma of Section 1.4). Thus, g moves points R which are arbitrarily close to Q, and the gR are equally close to $gQ = Q$ because g is an isometry. In other words, any neighborhood D_Q of Q includes distinct points R, gR in the same Γ-orbit, contrary to hypothesis.   □

We now appeal to the classification of euclidean isometries (Section 1.5) to determine the discontinuous fixed point free groups Γ. Since rotations and reflections have fixed points, Γ can include only translations or proper

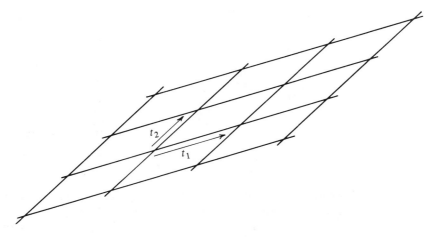

FIGURE 2.14.

glide reflections. The following theorem then establishes that Γ is one of the groups already considered in Sections 2.2 to 2.4.

Theorem. *A discontinuous, fixed point free group Γ of isometries of $\mathbb{R}^2$ is generated by one or two elements.*

Proof. First assume that Γ includes translations only. Choose any point $P \in \mathbb{R}^2$. By the discontinuity of Γ there is a minimum distance $\delta > 0$ to any other point in the Γ-orbit of P. We choose a translation t_1 of P to one of these nearest members of ΓP as the first generator of Γ.

The powers $\dots t_1^{-1}, 1, t_1, t_1^2, \dots$ of t_1 include all translations in Γ with the same direction as t_1. If $t \in \Gamma$ were another translation in the same direction, and if t_1^m sent P as near as possible to tP, then $t^{-1}t_1^m$ would be a translation in the same direction as t_1, but shorter, contrary to the choice of t_1.

Thus, if $\{\dots t_1^{-1}, 1, t_1, t_1^2, \dots\}$ does not exhaust Γ, the remaining translations have a different direction. In this case we choose from them a second generator t_2, again of minimal length. Since t_1 and t_2 are in different directions, they generate a *lattice* of points in $\mathbb{R}^2$—the vertices of a tessellation of $\mathbb{R}^2$ by equal parallelograms (Figure 2.14). It remains to show that this lattice is the whole Γ-orbit of P.

The argument is a generalization of that which shows that the powers of t_1^n exhaust all $t \in \Gamma$ with the same direction as t_1. We now have to show that if $t \in \Gamma$ is not a lattice translation, and if $t_1^m t_2^n$ sends P as near as possible to tP, then $t^{-1}t_1^m t_2^n$ is a translation shorter than t_1 or t_2, which is a contradiction.

This amounts to showing that any point Q within a parallelogram is separated from at least one vertex by a distance less than the length of the

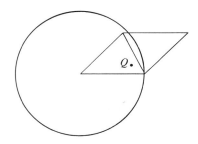

FIGURE 2.15.

longest side. In fact, since Q lies in one-half of the parallelogram, Q must lie inside a circle with one long side as radius (Figure 2.15).

This completes the proof that an isometry group Γ consisting only of translations has one or two generators. The proof is similar if Γ contains glide reflections because in this case it is still true that two elements generate a lattice (see Exercises 2.5.2 and 2.5.3).   □

Corollary. $S = \mathbb{R}^2/\Gamma$ *is a cylinder, twisted cylinder, torus, or Klein bottle.*

Proof. It follows from Sections 2.2 to 2.4 that we get the following surfaces, depending on the generators of Γ:

Surface	Generators of Γ
Cylinder	Single translation
Twisted cylinder	Single glide reflection
Torus	Two translations
Klein bottle	One translation, one glide reflection (or equivalently, as shown in Section 2.4, two glide reflections).

□

Exercises

2.5.1. Give a careful justification of the claim in the theorem that Q lies inside a circle with one long side as radius.

2.5.2. If Γ is a discontinuous, fixed point free group of glide reflections and translations, let g be a glide reflection of minimal length in Γ, and let h be an element of minimal length not in the direction of g. Show that g, h must have perpendicular directions (e.g., by finding shorter elements when the directions of g, h are not perpendicular).

2.5.3. Deduce from Exercise 2.5.2 and the method of the theorem of Section 2.5 that g, h generate Γ.

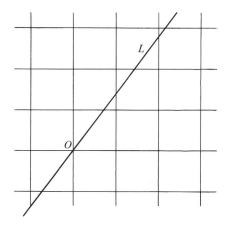

FIGURE 2.16.

2.6 A Nondiscontinuous Group

An unfortunate thing about the word "discontinuous", in the context of groups, is that the opposite is *not* "continuous". One simply has to say "nondiscontinuous". There are interesting nondiscontinuous groups associated with infinite lines on the torus.

Consider for simplicity the torus $T = \mathbb{R}^2/\Gamma$, where Γ is the group generated by unit translations in the x- and y-directions. Let L be a line through O on $\mathbb{R}^2$ with irrational slope, say $\sqrt{2}$. Figure 2.16 shows L on the tessellation of unit squares, each of which represents a copy of T when its opposite sides are identified. The orbit map $\Gamma\cdot$ maps L one-to-one into T because no two points of L are in the same Γ-orbit. [Two points in the same Γ-orbit are of the form (x, y) and $(x + m, y + n)$; so if both were on L, the slope of L would be the rational number n/m.]

Thus, the image $\Gamma L = \{\Gamma P \mid P \in L\}$ of L in T is an infinite line. I claim that ΓL *passes arbitrarily close to each point of* T.

This amounts to showing that the parallel lines gL for $g \in \Gamma$ fill $\mathbb{R}^2$ densely or, equivalently, that they meet the x-axis in a dense set of points. These points include 1 (by unit translation of L in the x-direction) and $1/\sqrt{2}$ (by unit translation of L in the y-direction). Therefore, since Γ is a group, they must include all points in the additive group $G \subset \mathbb{R}$ that is generated by 1 and $1/\sqrt{2}$. The claim is then justified by the following:

Theorem. *If α is any irrational real number, then the additive group $G \subset \mathbb{R}$ generated by 1 and α is dense in $\mathbb{R}$. That is, G includes points arbitrarily close to any point of $\mathbb{R}$.*

Proof. The infinitely many elements $\alpha, 2\alpha, 3\alpha, \ldots \in G$ are all distinct modulo 1, because if

$$ma - n\alpha = \text{an integer } p,$$

then

$$\alpha = \frac{p}{m - n},$$

contradicting the irrationality of α. Hence,

$$A = \{n\alpha - (\text{integer part of } n\alpha) \mid n = 1, 2, 3, \ldots\}$$

is an infinite subset of G in $[0, 1]$.

The infinitude of A implies that, for any $\epsilon > 0$, there are $\beta, \gamma \in A$ such that $|\beta - \gamma| < \epsilon$. Since G is a group, $\beta, \gamma \in G$ implies $\beta - \gamma \in G$ and also $p(\beta - \gamma) \in G$ for any integer p. The latter include points within ϵ of any real number. $\qquad\square$

The group G is a nondiscontinuous subgroup of $\mathbb{R}$ (or group of translations of $\mathbb{R}$) because the G-orbit of any point P includes points in any neighborhood of any point.

The group $G/\mathbb{Z}$, which is an infinite cyclic group generated by the coset $\{\alpha + n \mid n \in \mathbb{Z}\}$, is dense in $\mathbb{R}/\mathbb{Z}$, the circle obtained by identifying the ends 0 and 1 of the unit interval. The elements of $G/\mathbb{Z}$ can be viewed as rotations of the circle through integer multiples of $2\pi\alpha$. Thus, $G/\mathbb{Z}$ is an infinite cyclic, but dense (and hence nondiscontinuous) group of isometries of the circle.

Exercises

2.6.1. Show that the set of intersections of the lines gL with the x-axis is precisely the group G.

2.6.2. Suppose that each integer point $(m, n) \in \mathbb{R}^2$ is surrounded by a disc of radius $\epsilon > 0$. Show that, whatever the value of ϵ, each ray from O hits a disc. (The so-called *Olbers paradox* of astronomy is based on this argument.)

2.7 Euclidean Surfaces

As we have seen in Sections 2.2 to 2.5, each quotient surface $S = \mathbb{R}^2/\Gamma$ has a well-defined distance between any two points ΓP and ΓQ, namely,

$$d_S(\Gamma P, \Gamma Q) = \min\{d(P', Q') \mid P' \in \Gamma P,\ Q' \in \Gamma Q\},$$

where d denotes the (euclidean) distance in $\mathbb{R}^2$. The distance d_S is "locally euclidean" in the sense that each $\Gamma P \in S$ has an ϵ-neighborhood $D_\epsilon(\Gamma P) =$

$\{\Gamma Q \in S \mid d_S(\Gamma P, \Gamma Q) < \epsilon\}$ isometric to an ϵ-disc of $\mathbb{R}^2$, namely, the ϵ-disc $D_\epsilon(P)$ centered on P, where ϵ is less than one-quarter of the distance to the nearest $P' \neq P$ in the same Γ-orbit (cf. Section 2.2).

We now abstract this "locally euclidean" property in the following definition. A *euclidean surface* S is a set with a real-valued function $d_S(A, B)$ which is defined for all $A, B \in S$ and *locally euclidean* in the following sense. For each $A \in S$ there is an $\epsilon > 0$ such that the ϵ-*neighborhood of* A,

$$D_\epsilon(A) = \{B \in S \mid d_S(A, B) < \epsilon\},$$

is isometric to a euclidean disc. That is, there is a one-to-one correspondence $B \leftrightarrow B'$ between the points $B \in D_\epsilon(A)$ and the points B' of an open disc $D \subset \mathbb{R}^2$ such that

$$d_S(B_1, B_2) = d(B_1', B_2').$$

Thus, within such an ϵ-neighborhood, which we call a *euclidean disc of S*, the concepts of distance, angle, line segment, etc., have the usual euclidean meaning.

Locally euclidean surfaces can lack one geometrically natural property— possessed by the quotient surfaces $\mathbb{R}^2/\Gamma$ in particular—that of *connectedness*. For example, take S to be the union of the parallel planes $z = 0$ and $z = 1$ in $\mathbb{R}^3$, with

$$d_S(A, B) = \text{euclidean distance between } A, B \text{ in } \mathbb{R}^3.$$

Then for $\epsilon < 1$ the set $D_\epsilon(A)$ is indeed a euclidean disc, so S is a euclidean surface despite being obviously disconnected.

This example prompts us to restrict attention to euclidean surfaces which are "connected" in a reasonable sense. Intuitively, one wants to be able to move continuously between any two points on S, and this property can be formulated geometrically as follows: a euclidean surface S is *connected* if each $A, B \in S$ are connected by a polygonal path Σ whose sides lie within euclidean discs of S. Then quotient surfaces $\mathbb{R}^2/\Gamma$ are connected euclidean surfaces, whereas the union of parallel planes is not.

Finally, we impose a second condition on euclidean surfaces to exclude another type of pathology—incompleteness. The plane minus a point, for example, is a connected euclidean surface, but it is geometrically unsatisfactory because not every line segment can be continued indefinitely. On a euclidean surface S we can define a *line segment* to be a polygonal path whose successive sides (each lying within a euclidean disc of S where the notion of "line segment" is already meaningful) meet at angle π. Then S is called *complete* if any line segment can be continued indefinitely. It is easy to see that all quotient surfaces $\mathbb{R}^2/\Gamma$ are complete.

The mild restrictions of connectedness and completeness actually exclude all euclidean surfaces *except* the quotients $\mathbb{R}^2/\Gamma$. This far from obvious theorem is proved over the next two sections.

Remark. Although $\mathbb{R}^2 - \{O\}$ is incomplete under the usual euclidean distance in $\mathbb{R}^2$, it can be made complete (and even euclidean) by a different choice of distance function. See Exercises 2.7.3 and 2.7.4, and also Section 5.3. Such a function naturally assigns "infinite length" to any path in $\mathbb{R}^2$ which approaches O.

Exercises

2.7.1. For each of the following "surfaces" (assuming the obvious local distance function) state whether it is euclidean, connected, or complete.

  (i) The union of the planes $x = 0$ and $y = 0$ in $\mathbb{R}^3$.

  (ii) The surface of an infinite triangular prism.

  (iii) The surface of a cube.

  (iv) The surface of a cube, minus the vertices.

  (v) A closed disc.

  (vi) An open disc.

2.7.2. If $P_1, P_2, \ldots \in \mathbb{R}^2$, show that $\mathbb{R}^2 - \{P_1, \ldots, P_n\}$ is a euclidean surface but that $\mathbb{R}^2 - \{P_1, P_2, \ldots\}$ (for certain $P_1, P_2, \ldots$) is not.

2.7.3. Show that the exponential function maps the strip $-\pi \leq \mathrm{Im}(z) \leq \pi$ of $\mathbb{C}$ onto $\mathbb{C} - \{O\}$, giving a homeomorphism of the cylinder onto $\mathbb{C} - \{O\}$.

2.7.4. Deduce that the minimum of the infinitely many values, for fixed z_1, z_2, of $|\log(z_1/z_2)|$ is a complete locally euclidean distance function on $\mathbb{C} - \{O\}$, with isometries $r_\theta(z) = e^{i\theta}z$ and $d_\rho(z) = \rho z$ for $\theta, \rho \in \mathbb{R}$. Interpret r_θ, d_ρ on the cylinder.

2.8   Covering a Surface by the Plane

The key to the representation of a complete, connected euclidean surface S as a quotient $\mathbb{R}^2/\Gamma$ is the construction of a certain map $\mathbb{R}^2 \to S$ called a *covering*. Like an orbit map $\mathbb{R}^2 \to \mathbb{R}^2/\Gamma$, but conceivably more general, a covering is a map f of $\mathbb{R}^2$ onto S which is a *local isometry*. That is, each $P \in \mathbb{R}^2$ has an ϵ-neighborhood $D_\epsilon(P)$ in which $d_S(f(Q), f(R)) = d(Q, R)$ for all Q, R.

Since S, by definition, is locally like $\mathbb{R}^2$, it is plausible that we can "wrap" $\mathbb{R}^2$ around S a small piece at a time to achieve a covering. However, the details are quite delicate and they depend crucially on the connectedness and completeness of S. The first rigorous proof that $\mathbb{R}^2$ covers S was given by Hopf [1925], using a device we shall call the *pencil map*.

The pencil map exploits the fact that $\mathbb{R}^2$ is filled by the family of rays with origin O (the O-*pencil*), so that each point $P \in \mathbb{R}^2$ is uniquely determined by the ray from O through P and the length $|OP|$ of OP. We choose a point $O^S \in S$ and an isometry $p : D_\epsilon(O) \to D_\epsilon(O^S)$, which is possible for sufficiently small ϵ since S is a euclidean surface. The *pencil map* $p : \mathbb{R}^2 \to S$ is simply the natural extension of this isometry along the rays of the O-pencil: $p(P)$ is found by extending the line segment out of O^S which is the p-image of $OP \cap D_\epsilon(O)$ to distance $|OP|$.

Since S is complete, any line segment on it can be extended indefinitely, hence the pencil map is defined for each $P \in \mathbb{R}^2$. Of course, many points of $\mathbb{R}^2$ may have the same p-image, as we can already anticipate from the example of the covering of the cylinder (Section 2.2). The following theorem shows that the pencil map is a covering as defined above.

Theorem. *The pencil map p has the properties*

(i) *each $P \in \mathbb{R}^2$ has a neighborhood on which p is an isometry, and*

(ii) *p is onto S.*

To prove this theorem we shall appeal to the compactness of closed line segments in $\mathbb{R}^2$. The first form of compactness we shall use is: a nonempty closed subset of a closed line segment has a least member (where "least" means nearest to a specified end of the segment). The second is the *Heine–Borel* theorem: if a closed line segment L is contained in a union of (infinitely many) open discs, then the union of finitely many of these discs also contains L. We assume that the reader knows the basic facts about open and closed sets, but some exercises are offered for those who want to refresh their memories about compactness.

Proof of the Theorem. (i) Suppose on the contrary that $P \in \mathbb{R}^2$ has no neighborhood on which p is an isometry. We say that this is a point where p is *not locally isometric*. Such points form a closed set because if Q has a neighborhood on which p *is* an isometry, then so have all points in this neighborhood. Thus, the points on line segment OP at which p is not locally isometric form a nonempty closed set, and hence have a least member. Without loss of generality we can assume that this least member (the one nearest to O) is P itself.

Since $p(P)$ lies in the euclidean surface S, it has an ϵ-neighborhood $D_\epsilon(p(P))$ isometric to a disc of $\mathbb{R}^2$, which we can choose to be $D_\epsilon(P)$. We can rotate $D_\epsilon(P)$ to make the isometry agree with p on $OP \cap D_\epsilon(P)$. Finally, we can reflect in OP, if necessary, to make the isometry agree with p on any subdisc of $D_\epsilon(P)$ centered on OP, where p is itself an isometry. By hypothesis, p is indeed an isometry on a sufficiently small neighborhood $D_\delta(Q)$ of any Q in $D_\epsilon(P)$ between O and P (Figure 2.17).

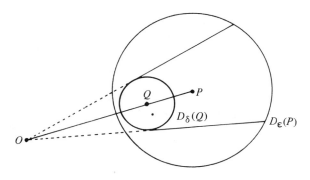

FIGURE 2.17.

We now have two maps, an isometry $f : D_\epsilon(P) \rightarrow D_\epsilon(p(P))$ and the pencil map $p : D_\epsilon(P) \rightarrow D_\epsilon(p(P))$, which agree on $D_\delta(Q)$. But f and p *also* agree on the prolongations of all rays from 0 which pass through $D_\delta(Q)$ (Figure 2.17) because f preserves length and straightness. The union of these line segments includes a disc neighborhood of P, so we have a contradiction.

(ii) To prove that p is onto S we first note that p maps any closed line segment L of $\mathbb{R}^2$ onto a line segment of S. By compactness and the local isometry of p, L can be divided into subsegments, each lying in a disc of $\mathbb{R}^2$ which is mapped isometrically by p. Then $p(L)$ is a union of line segments of S, each meeting its predecessor at angle π (as do the preimage segments in $\mathbb{R}^2$); hence $p(L)$ is a line segment of S.

Now if p is not onto S, consider the nonempty set $S - p(\mathbb{R}^2)$. Since p is a local isometry, $p(\mathbb{R}^2)$ includes an ϵ-neighborhood of each $p(P)$, and hence is open, so $S - p(\mathbb{R}^2)$ is closed. If $P^S \in S - p(\mathbb{R}^2)$, connectedness of S gives a polygonal path from O^S to P^S, and then there is a first point Q^S of this path in $S - p(\mathbb{R}^2)$ (Figure 2.18). If R^S is the last vertex of this path in $p(\mathbb{R}^2)$, let $R^S = p(R)$ and consider any line segment L out of R which is mapped by p onto an initial segment of $R^S Q^S$. By prolonging L sufficiently we can make its p-image include all of $R^S Q^S \cap p(\mathbb{R}^2)$ and hence, by continuity, it will include Q^S also, which is a contradiction.   □

Remark. Part (ii) of the theorem would be immediate if we knew that each $P^S \in S$ lay on a ray through O^S. This is indeed the case (in fact more is true, see Exercise 2.8.3) but it does not seem possible to prove it without proving (ii) first.

Exercises

2.8.1. Assume that $\mathbb{R}$ (and hence any line) has the following *completeness* property: any bounded set $B \subset \mathbb{R}$ has a least upper bound.

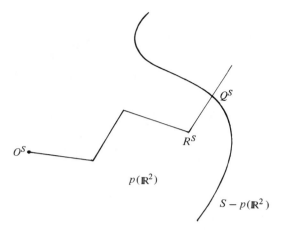

FIGURE 2.18.

(i) Prove that any nonempty closed subset K of an interval $[a, b]$ has a least member by considering

$$B = \{c \in [a, b] \mid K \cap [a, c] = \emptyset\}.$$

(ii) Prove that if $[a, b]$ is contained in a union of (infinitely many) open intervals, then the union of some finite subset of the intervals also contains $[a, b]$ by considering

$$B = \{c \in [a, b] \mid [a, c] \subset \text{ the union of finitely many intervals}\}.$$

2.8.2. Deduce the properties of line intervals in $\mathbb{R}^2$ used in the proof of the theorem above.

2.8.3. Show that any two points of a complete euclidean surface may be connected by a line segment (the p-image of a line segment in $\mathbb{R}^2$).

2.8.4. Give an appropriate definition of a covering of one euclidean surface by another, and hence show that a cylinder covers the twisted cylinder, and a torus covers the Klein bottle.

2.9 The Covering Isometry Group

The construction of a covering $p : \mathbb{R}^2 \to S$ for a complete, connected euclidean surface S is a big step towards the representation of S as a quotient $\mathbb{R}^2/\Gamma$. What remains to be seen is that p is an orbit map because the group Γ is still not evident. It materializes as the *covering isometry group* of p.

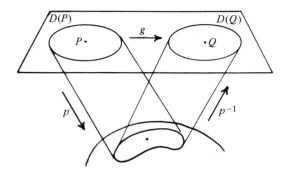

FIGURE 2.19.

An isometry $g : \mathbb{R}^2 \to \mathbb{R}^2$ is called a *covering isometry* (for a covering $f : \mathbb{R}^2 \to S$) if $fgP = fP$ for all $P \in \mathbb{R}^2$. In other words, for all $P \in \mathbb{R}^2$, the points P and gP "lie over" the same point of S. The covering isometries form a group Γ. It is immediate from the definition that if g_1, g_2 are covering isometries, then so is $g_1 g_2$. Also, since an isometry g is invertible, we can write any $P \in \mathbb{R}^2$ as $g^{-1}Q$, so the covering isometry condition becomes $fgg^{-1}Q = fg^{-1}Q$, i.e., $fQ = fg^{-1}Q$ for all $Q \in \mathbb{R}^2$, which says that g^{-1} is also a covering isometry.

Thus, if Γ is the covering isometry group for p, the whole Γ-orbit of a point P is mapped by p to the same point of S. To show that p is, in fact, the Γ-orbit map, it remains to show the following.

Theorem. *If $pP = pQ$ then $Q = gP$ for some covering isometry g (i.e., P, Q are in the same Γ-orbit).*

Proof. By the local isometry property of p, there are disc neighborhoods $D(P)$ of P and $D(Q)$ of Q mapped isometrically by p onto a disc $pD(P) = pD(Q)$ of S (Figure 2.19). Thus, $D(P) \overset{p}{\longrightarrow} pD(P) \overset{p^{-1}}{\longrightarrow} D(Q)$ is an isometry $g : D(P) \to D(Q)$ and because $D(P)$ contains three points not in a line it extends uniquely to an isometry $g : \mathbb{R}^2 \to \mathbb{R}^2$ by the lemma in Section 1.4. We wish to show that g is a covering isometry for p, i.e., that $pR = pgR$ for all $R \in \mathbb{R}^2$.

As in Section 2.8 we can argue that if $pR \neq pgR$ for any R, there is such an R which is "least", i.e., nearest to P on some line through P. The reason is that the set $\{R \mid pR = pgR\}$ is open, by the argument just given to show that p and pg agree on a neighborhood of P, and hence $\{R \mid pR \neq pgR\}$ is closed.

Given that R is the nearest point to P on PR such that $pR \neq pgR$, consider a sequence $R_1, R_2, R_3, \ldots$ of points between P and R which converge to R. Then by hypothesis

$$pR_i = pgR_i,$$

and by continuity of p and g,

$$p\left(\lim_{i\to\infty} R_i\right) = pg\left(\lim_{i\to\infty} R_i\right);$$

that is,

$$pR = pgR,$$

and we have a contradiction.

Hence, $pR = pgR$ for all $R \in \mathbb{R}^2$, and hence g is a covering isometry. $\square$

Corollary (Killing–Hopf theorem). *Each complete, connected euclidean surface is of the form $\mathbb{R}^2/\Gamma$, and hence is either a cylinder, twisted cylinder, torus, or Klein bottle (if not $\mathbb{R}^2$ itself).*

Proof. This is immediate from the theorem, the theorem of Section 2.8, and the corollary of Section 2.5. $\square$

2.10   Discussion

The Killing–Hopf theorem is much more than a theorem about euclidean surfaces. Since the proof depends mainly on the unique determination of points by direction and distance from the origin, it applies equally to euclidean spaces of any dimension. It also applies, as we shall see, to hyperbolic spaces and to spheres (only in the latter case requiring slight modification). The general conclusion is that any space of constant curvature is the quotient of euclidean, hyperbolic, or spherical space by a discontinuous, fixed point free group Γ of isometries.

The concept of "covering" which emerges in the Killing–Hopf theorem is actually more special than is usual, in two respects. First, it is usual to require only local homeomorphism rather than local isometry; the usual covering concept is topological rather than geometric. In our case, the local isometry condition is not a significant specialization because we *define* distance on the quotient surface so that local isometry is automatic. But our coverings really are special in a second respect, inasmuch as the covering surface is always the euclidean plane. It is quite possible to define coverings by other surfaces. For example, a torus can be covered by a cylinder. The plane enjoys the distinction of being able to cover any euclidean surface, and for that reason it is called the *universal* covering surface.

The concept of universal covering is usually developed in the topological setting, where it is shown that any reasonable space has a universal covering. The universal covering of S covers any space which covers S. The topological construction of the universal covering is quite different from the Hopf construction we used in Section 2.8, and it yields new insight into the group Γ. We shall study this construction in Chapter 6.

FIGURE 2.20.

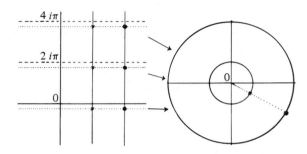

FIGURE 2.21.

The covering of the cylinder by the plane has been known in some sense since ancient times. It was realized artistically in the *cylinder seals* of Mesopotamia from c. 3000 BC. A picture carved on a cylinder is imprinted periodically on the plane by rolling the cylinder on a soft clay surface (Figure 2.20).

When we interpret the euclidean plane as $\mathbb{C}$, universal coverings can be realized by natural complex functions. The exponential function $f(z) = e^z$, for example, maps $\mathbb{C}$ onto $\mathbb{C} - \{0\}$, which is topologically a cylinder (cf. Exercise 2.7.3). This depends on the miraculous fact that

$$e^{x+iy} = e^x(\cos y + i \sin y),$$

so that successive strips $\ldots, 0 \leq y \leq 2\pi, 2\pi \leq y \leq 4\pi, \ldots$ of $\mathbb{C}$ are wrapped round and round $\mathbb{C}-\{0\}$ (Figure 2.21). More precisely, e^z takes equal values at all points in an orbit ΓP, where Γ is the group generated by $g(z) = 2i\pi+z$ (two such orbits, and their images, are shown in Figure 2.21). Consequently, e^z is well-defined on the orbit space $\mathbb{C}/\Gamma$, which is, of course, a cylinder. The orbits are in one-to-one continuous correspondence with the points e^z of $\mathbb{C} - \{0\}$, which again shows that $\mathbb{C} - \{0\}$ is topologically a cylinder.

All this is implicit in the discoveries of Euler [1748] relating complex exponentials and logarithms to the circular functions. (It should be stressed, however, that the idea of complex functions as mappings of the plane was still hazy in Euler's time. The first clear understanding was shown by Gauss [1811], whose analysis of the complex logarithm will be taken up when we discuss the evolution of the topological covering concept in Chapter 6.)

Euler's book [1748, p. 191], also contains the first formula for a circular function which makes its periodicity obvious, namely,

$$\pi \cot \pi z = \sum_{n=-\infty}^{\infty} \frac{1}{z+n}. \tag{1}$$

Periodicity is formally obvious because replacement of z by $z+1$ gives the same set of terms on the right-hand side. One only has to be careful about the meaning of the sum to ensure convergence. Eisenstein [1847] considered the analogous formula

$$f(z) = \sum_{m,n=-\infty}^{\infty} \frac{1}{(z+m+ni)^2}, \tag{2}$$

for which it is equally clear that f is unchanged when z is replaced by $z+1$ or $z+i$ (the terms are squared to ensure convergence, which again requires some thought). Thus, Eisenstein's function has *two* periods, 1 and i, and it is well-defined on the orbit space $\mathbb{C}/\Gamma$, where Γ is the group generated by $g(z) = 1+z$ and $h(z) = i+z$. This orbit space is a torus.

To view the *range* of f as a torus, one has to regard the f-values of distinct Γ-orbits as distinct—a ticklish problem because, in general, f maps *two* orbits to the same complex number [because $1/(-z)^2 = 1/z^2$]. There is also the problem that $f(z) = \infty$ for z in the Γ-orbit of 0. Enlarging $\mathbb{C}$ to $\mathbb{C} \cup \{\infty\}$ is actually not difficult. The result is a sphere, as we will see in Chapter 3. "Counting the same complex number more than once" is the more delicate construction of *Riemann surfaces* or *branched coverings of the sphere*, which we will meet in Chapter 8.

Doubly periodic functions such as (2) are called *elliptic functions* because an early example arose from the formula for the arc length of an ellipse. This formula is actually a (multiple-valued) integral rather than a function, and the early development of the subject was laborious because it focused on elliptic integrals rather than on their inverses, which are elliptic functions. The theory of elliptic functions finally took off in the 1820s, when Abel and Jacobi realized that inversion of elliptic integrals makes things easier (just as the exponential function is easier than the logarithm, and the sine function is easier than the arcsine integral). They also recognized that elliptic functions should be viewed as functions of a complex variable because only then does their key property of double periodicity emerge.

A sketch of the history of elliptic functions, plus references to more detailed information, may be found in Stillwell [1989].

The Eisenstein formula (2) has a more general form

$$f(z) = \sum_{m,n=-\infty}^{\infty} \frac{1}{(z+m\omega_1+n\omega_2)^2}, \tag{3}$$

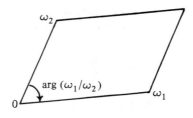

FIGURE 2.22.

where ω_1, ω_2 are any two complex numbers. Provided $\omega_1/\omega_2 \notin \mathbb{R}$, so that ω_1, ω_2 have different directions, $f(z)$ has two basic periods ω_1 and ω_2. Thus, for any torus $\mathbb{C}/\Gamma$ there is an elliptic function which "belongs to" $\mathbb{C}/\Gamma$. Much of the geometry and topology of tori was, in fact, worked out in the 19th century to support the theory of elliptic functions. For example, the dependence of (3) on ω_1, ω_2 is reflected in the "shape" ω_1/ω_2 of the fundamental region for Γ shown in Figure 2.22. This focuses attention on the space of "shapes of parallelograms" which, rather surprisingly, turns out to be a hyperbolic plane (see Chapters 4, 5, and 6).

All this shows that cylinders and tori have a natural place in the theory of complex functions. Do the twisted cylinder and the Klein bottle also belong there? No, in this respect the nonorientable surfaces are definitely second class citizens. Their entry is barred by the implicit assumption of differentiability. A differentiable complex function preserves the sign as well as the magnitude of angles. Consequently, such a function can only map $\mathbb{C}$ onto an orientable surface. In particular, it is not possible for complex functions to realize the orbit maps from the euclidean plane to the twisted cylinder and the Klein bottle.

3

The Sphere

3.1 The Sphere $\mathbb{S}^2$ in $\mathbb{R}^3$

In classical geometry the sphere is viewed as a figure in three-dimensional euclidean space, analogous to the circle in the euclidean plane. The circle, however, is of interest *only* in relation to the plane. Its intrinsic structure is locally the same as the line because we have the map $\theta \mapsto e^{i\theta}$ which is a local isometry between the line and the unit circle. The sphere, on the other hand, is *not* locally isometric to the plane, hence it is of interest as a self-contained structure. This intrinsic structure makes the sphere the first example of a non-euclidean geometry.

As it happens, all the intrinsic properties of the sphere $\mathbb{S}^2$ are inherited from the surrounding euclidean space $\mathbb{R}^3$. For example, the isometries of $\mathbb{S}^2$ turn out to be simply the isometries of $\mathbb{R}^3$ that leave O fixed. Thus, the noneuclidean geometry of $\mathbb{S}^2$ can, in fact, be described in terms of the euclidean geometry of $\mathbb{R}^3$. We shall start with this point of view, but later pass to the intrinsic viewpoint by mapping $\mathbb{S}^2$ (nonisometrically) onto the euclidean plane, which we take to be the plane $\mathbb{C}$ of complex numbers. This reveals a striking connection between the geometry of $\mathbb{S}^2$ and the algebra of $\mathbb{C}$. The algebra involved (linear fractional transformations) also turns out to be the key to our second example of non-euclidean geometry—the hyperbolic plane—which will be studied in Chapter 4.

The geometry of $\mathbb{R}^3$, like that of $\mathbb{R}^2$, is based on the Pythagorean distance formula: the distance between points $P_1 = (x_1, y_1, z_1)$ and $P_2 = (x_2, y_2, z_2)$ is

$$d(P_1, P_2) = \sqrt{(x_2 - x_1)^2 + (y_2 - y_1)^2 + (z_2 - z_1)^2}.$$

A *plane* is given by a linear equation

$$\alpha x + \beta y + \gamma z = \delta$$

and a *line* is given by the intersection of suitable planes or, more conveniently, by linear equations in a parameter t:

$$x = x_0 + \lambda t, \ y = y_0 + \mu t, \ z = z_0 + \nu t.$$

The *unit sphere* $\mathbb{S}^2$ consists of all points (x, y, z) at distance 1 from O, and hence has equation

$$x^2 + y^2 + z^2 = 1.$$

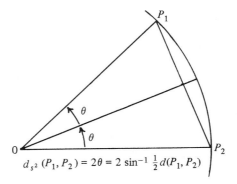

FIGURE 3.1.

The *spherical distance* $d_{\mathbb{S}^2}(P_1, P_2)$ between P_1, $P_2 \in \mathbb{S}^2$ is seen from Figure 3.1 to be given by

$$d_{\mathbb{S}^2}(P_1, P_2) = 2\theta = 2\sin^{-1} \tfrac{1}{2} d(P_1, P_2).$$

Since $d_{\mathbb{S}^2}(P_1, P_2)$ is a function of $d(P_1, P_2)$, "spherically equidistant" points are also equidistant in $\mathbb{R}^3$, which means that in many situations we can use $d(P_1, P_2)$ instead of the more complicated $d_{\mathbb{S}^2}(P_1, P_2)$.

We shall now find all the isometries of $\mathbb{S}^2$, much as we did for $\mathbb{R}^2$ in Section 1.4, by finding enough isometries of $\mathbb{R}^3$ to map any triangle on $\mathbb{S}^2$ to any isometric triangle on $\mathbb{S}^2$.

Certain isometries of $\mathbb{R}^3$ are obvious, being essentially isometries of $\mathbb{R}^2$ with a redundant coordinate attached. For example, *rotation $r_{z,\theta}$ about the z-axis by θ* is given by

$$x' = x \cos\theta - y \sin\theta,$$
$$y' = x \sin\theta + y \cos\theta,$$
$$z' = z$$

and the check that this leaves $d(P_1, P_2)$ unchanged is essentially the same as for $\mathbb{R}^2$. Similarly, for *reflection $\bar{r}_E$ in the plane $z = 0$*:

$$x' = x,$$
$$y' = y,$$
$$z' = -z.$$

(Anticipating the notation $\bar{r}_L$ for reflection in a "line" L, we call this reflection $\bar{r}_E$ because it is reflection in the equator.)

Using the rotations $r_{x,\theta}$, $r_{y,\theta}$ (analogous to $r_{z,\theta}$, but about the x- and y-axes respectively), one can move any line through O to the position of the x-axis. Then, by further rotation about the x-axis one can bring any plane through O to the position of the plane $z = 0$. (Rotations, being linear transformations, do indeed map planes to planes.)

It is, therefore, valid to treat any line through O as a coordinate axis, and any plane through O as a coordinate plane. In particular, we can define reflection in an arbitrary plane Π by conjugating $\bar{r}_E$ with a sequence of rotations bringing Π to the position of the plane $z = 0$.

Analogous to the equidistant property of a line in $\mathbb{R}^2$, we have the following in $\mathbb{R}^3$:

Lemma. *The set of points equidistant from two points $P, P' \in \mathbb{S}^2$ is a plane through O, and reflection in this plane exchanges P and P'.*

Proof. Let $P = (\alpha, \beta, \gamma)$ and $P' = (\alpha', \beta', \gamma')$. If (x, y, z) is equidistant from P and P', we have

$$(x - \alpha)^2 + (y - \beta)^2 + (z - \gamma)^2 = (x - \alpha')^2 + (y - \beta')^2 + (z - \gamma')^2.$$

Since $\alpha^2 + \beta^2 + \gamma^2 = \alpha'^2 + \beta'^2 + \gamma'^2 = 1$ by the assumption that $P, P' \in \mathbb{S}^2$, the equation for (x, y, z) simplifies to

$$-2\alpha x - 2\beta y - 2\gamma z = -2\alpha' x - 2\beta' y - 2\gamma' z$$

or

$$(\alpha - \alpha')x + (\beta - \beta')y + (\gamma - \gamma')z = 0, \tag{1}$$

which is the equation of a plane through O.

If we choose axes so that P, P' lie in $z = 0$ and are mirror images in the x-axis, $P = (\alpha, \beta, 0)$, $P' = (\alpha, -\beta, 0)$, then the equidistant plane (1) becomes $y = 0$, reflection in which does indeed exchange P and P'.   □

Thus if we define a *line* of $\mathbb{S}^2$ to be the equidistant set of two points $P, P' \in \mathbb{S}^2$, a line is the intersection of $\mathbb{S}^2$ with a plane through O, i.e., a *great circle*. Reflections of $\mathbb{S}^2$ can then be regarded as reflections in lines, and we have a three reflections theorem for $\mathbb{S}^2$ analogous to the one for $\mathbb{R}^2$.

Theorem. *Any isometry of $\mathbb{S}^2$ is the product of one, two, or three reflections.*

Proof. As in Section 1.4, one first proves a lemma that an isometry f of $\mathbb{S}^2$ is determined by the images of three points not in a line. Thanks to the lemma above, the proof is exactly the same as before, given the new interpretation of "line."

The theorem follows from the lemma in exactly the same way as for $\mathbb{R}^2$.   □

Corollary. *The isometries of $\mathbb{S}^2$ form a group, $\mathrm{Iso}(\mathbb{S}^2)$.*

Proof. Each reflection $\bar{r}_C$ in a great circle C is self-inverse; hence a product $\bar{r}_{C_1} \dots \bar{r}_{C_n}$ has an inverse, namely, $\bar{r}_{C_n} \dots \bar{r}_{C_1}$. The associative and identity properties are obvious.   □

Exercises

3.1.1. Show that two great circles intersect in a pair of *antipodal* points, i.e., a pair of the form (x, y, z), $(-x, -y, -z)$.

3.1.2. Show that antipodal points remain antipodal under any isometry of $\mathbb{S}^2$.

3.1.3. Using properties of lines, or otherwise, show that $\mathbb{S}^2$ is not a euclidean surface.

3.2   Rotations

Pursuing the analogy between $\mathbb{R}^2$ and $\mathbb{S}^2$ a little further, we should expect the rotations of $\mathbb{S}^2$ to be products of two reflections. This is certainly true of rotations $r_{z,\theta}$ about the z-axis because they are just rotations r_θ of the (x, y)-plane with a z-coordinate added. The representation of r_θ as a product of reflections in lines through O at angles 0 and $\theta/2$ gives a representation of $r_{z,\theta}$ as a product of reflections in planes through the z-axis at angles 0 and $\theta/2$ (Figure 3.2). Moreover, *any* planes Π, Π' through the z-axis give the same result, provided the angle between them is $\theta/2$ and in the same sense. This can be seen, as in Section 1.3, by conjugating each reflection by $r_{z,\phi}$, so as to rotate the planes of reflection through an arbitrary angle ϕ.

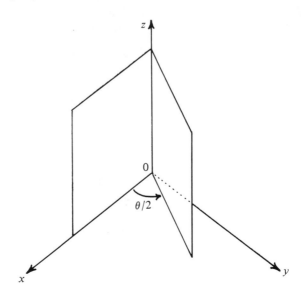

FIGURE 3.2.

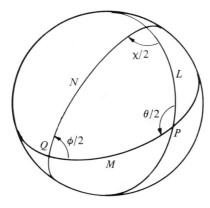

FIGURE 3.3.

Using the freedom to interpret any line K through O as the z-axis, this fact can be stated as the following general lemma: *a rotation $r_{K,\theta}$ about any line K through angle θ is the product of reflections in any planes Π, Π' through K separated by angle $\theta/2$*. To express this lemma in intrinsically, spherical terms we replace K by one of its intersections P with $\mathbb{S}^2$, and Π, Π' by the lines, i.e., great circles, L, L' in which they intersect $\mathbb{S}^2$. The statement then becomes: *a rotation $r_{P,\theta}$ of $\mathbb{S}^2$ about P through angle θ is the product of reflections in any lines L, L' through P separated by angle $\theta/2$*.

This elementary lemma, combined with the fact that any two lines on $\mathbb{S}^2$ intersect, has the following important consequence:

Theorem. *The product of two rotations of $\mathbb{S}^2$ is also a rotation. The rotations of $\mathbb{S}^2$ form a group.*

Proof. (Cf. Corollary 2 of Section 1.3). Given a rotation $r_{P,\theta}$ about P and a rotation $r_{Q,\phi}$ about Q, we choose lines of reflection L, M, N as shown in Figure 3.3. Denoting the reflections in these lines by $\bar{r}_L$, $\bar{r}_M$, $\bar{r}_N$, respectively, we have

$$r_{P,\theta} = \bar{r}_M \bar{r}_L, \quad r_{Q,\phi} = \bar{r}_N \bar{r}_M$$

and hence

$$r_{Q,\phi} r_{P,\theta} = \bar{r}_N \bar{r}_M \bar{r}_M \bar{r}_L = \bar{r}_N \bar{r}_L = r_{R,\chi},$$

where χ is measured in the sense shown in Figure 3.3.

Thus, the set of rotations of $\mathbb{S}^2$ is closed under product. It is also closed under inverse because $r_{P,\theta}^{-1} = r_{P,-\theta}$, and hence is a group. $\qquad\square$

The theorem shows that the rotations make up the group $\mathrm{Iso}^+(\mathbb{S}^2)$ of *orientation-preserving* isometries, defined as those that are products of an even number of reflections. These are complemented by the *orientation-reversing* isometries—the products of an odd number of reflections—which make up the coset $\mathrm{Iso}^+(\mathbb{S}^2) \cdot \bar{r}_L$ for any reflection $\bar{r}_L$. As in the case of $\mathbb{R}^2$, a fixed point argument shows that $\bar{r}_L \notin \mathrm{Iso}^+(\mathbb{S}^2)$: the fixed point set of a reflection is a line, whereas the fixed point set of a nontrivial rotation is a pair of points.

The difference between orientation-preserving and orientation-reversing isometries also shows up clearly when they are represented by complex functions. The orientation-preserving isometries are linear fractional functions of w (we now call the complex variable w to avoid confusion with the third coordinate in $\mathbb{R}^3$), whereas the orientation-reversing isometries are linear fractional functions of $\bar{w}$. This representation will be constructed in the next two sections.

Exercises

3.2.1. If r, r' are rotations of $\mathbb{S}^2$, under what circumstances does $rr' = r'r$?

3.2.2. Find a product of three reflections with no fixed points.

3.3   Stereographic Projection

It is not possible to map the whole sphere bijectively and continuously onto the plane; however, *stereographic projection* (Figure 3.4) maps all but one point of $\mathbb{S}^2$, the "north pole" $N = (0, 0, 1)$, which is the center of projection. If we let u, v be the x, y coordinates of the image point P' of $P = (x, y, z)$ under this projection, then the line NP' has parametric equations

$$x = 0 + tu, \quad y = 0 + tv, \quad z = 1 - t. \tag{1}$$

Intersecting this line with $\mathbb{S}^2$, whose equation is

$$x^2 + y^2 + z^2 = 1$$

we get $t^2u^2 + t^2v^2 + (1-t)^2 = 1$ i.e., $t^2(u^2 + v^2 + 1) = 2t$, from which $t = 0$ corresponding to N) or $t = 2/(u^2 + v^2 + 1)$. Substituting the latter value in (1), we get

$$(x, y, z) = \left(\frac{2u}{u^2 + v^2 + 1}, \frac{2v}{u^2 + v^2 + 1}, \frac{u^2 + v^2 - 1}{u^2 + v^2 + 1} \right). \tag{2}$$

Conversely, solving (1) for u, v in terms of x, y, z, we find

$$u = \frac{x}{t} = \frac{x}{1 - z},$$

$$v = \frac{y}{t} = \frac{y}{1 - z}. \tag{3}$$

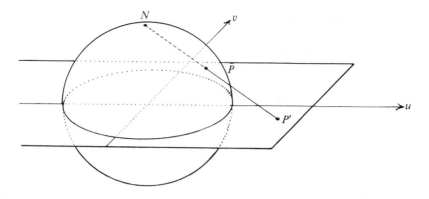

FIGURE 3.4.

In other words, stereographic projection sends $(x, y, z) \in \mathbb{S}^2$ to $\left(\frac{x}{1-z}, \frac{y}{1-z}\right)$ in the xy-plane, and the inverse map sends (u, v) in the xy-plane to

$$\left(\frac{2u}{u^2 + v^2 + 1}, \frac{2v}{u^2 + v^2 + 1}, \frac{u^2 + v^2 - 1}{u^2 + v^2 + 1}\right) \in \mathbb{S}^2.$$

The fact that the point $N = (0, 0, 1) \in \mathbb{S}^2$ does not correspond to a point of the plane is far from being a disadvantage of this mapping. In fact, it prompts us to view the plane as a plane of numbers (the complex numbers, naturally) and to "complete" the plane by the number ∞ corresponding to the point N. The clinching argument in favor of this idea is that the isometries of $\mathbb{S}^2$ then correspond to simple complex functions in which the number ∞ plays a necessary role (see Sections 3.4 and 3.5).

Stereographic projection is obviously not an isometry because distances are bounded on the sphere but unbounded in the plane. Nevertheless, it does preserve other geometric properties, most notably the angle between curves. The latter property—of *conformality* or "similarity in the small"— is the underlying reason why complex functions show up in the geometry of the sphere. As readers familiar with complex analysis will know, once any surface has been provided with a complex coordinate, the conformal mappings of the surface correspond to complex functions. We shall not prove this conformal property of stereographic projection here, as it will be easier once we have a complex coordinate on the sphere and representation of isometries by complex functions.

Exercises

3.3.1. Show that stereographic projection preserves angles at the "south pole" $(0, 0, -1)$ of $\mathbb{S}^2$. (Hence, angles are preserved at any point assuming that isometries of the sphere correspond to conformal mappings of the plane.)

3.3.2. Stereographic projection is often defined by projecting from the "north pole" onto the tangent plane $z = -1$ at the "south pole". What difference does it make?

3.4 Inversion and the Complex Coordinate on the Sphere

The link between isometries of $\mathbb{S}^2$ and complex functions is a geometric transformation known as *inversion*. If C is a circle in the euclidean plane and P is a point not at the center of C, then the *inverse* of P *with respect to* C is the point $\bar{P}_C$ shown in Figure 3.5. $\bar{P}_C$ lies on the line OP from the center O of C, at distance $|O\bar{P}_C|$ satisfying

$$|O\bar{P}_C| \cdot |OP| = \rho^2,$$

where ρ is the radius of C. Euclidean reflection in a line is a limiting case of inversion, obtained by letting $\rho \to \infty$ by moving O, while keeping the intersection of OP with C fixed. In fact, it is convenient to include euclidean reflections among the inversions by definition (regarding lines as "circles of infinite radius") to avoid continual mention of reflection as an exceptional case in the statement of theorems.

Thus, we shall define an *inversion* to be either a euclidean reflection $\bar{r}_L$ in a line L, or a function $I_C(P) = \bar{P}_C$ sending each point P to its inverse $\bar{P}_C$ with respect to a circle C.

It is not difficult to show that, when the plane is taken to be the plane $\mathbb{C}$ of complex numbers w, inversions are simple complex functions of $\bar{w}$. We shall work out the form of these functions below. But first we shall establish the connection with isometries of $\mathbb{S}^2$.

Theorem. *If* $\mathbb{S}^2 \subset \mathbb{R}^3$ *is reflected in a plane through* O, *then the induced map of stereographic images is an inversion.*

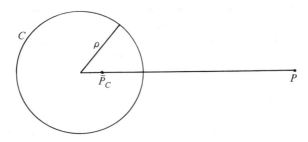

FIGURE 3.5.

Proof. Let $P = (\alpha, \beta, \gamma)$ and $P' = (\alpha', \beta', \gamma')$ be two points of $\mathbb{S}^2$ exchanged by the reflection, so the plane of reflection is

$$(\alpha - \alpha')x + (\beta - \beta')y + (\gamma - \gamma')z = 0 \tag{1}$$

as found in Section 3.1. The points (x, y, z) on this plane which are also on $\mathbb{S}^2$ are

$$\left(\frac{2u}{u^2 + v^2 + 1}, \frac{2v}{u^2 + v^2 + 1}, \frac{u^2 + v^2 - 1}{u^2 + v^2 + 1} \right)$$

in terms of their coordinates u, v on the stereographic image plane; hence the u, v on the stereographic image of the great circle of reflection satisfy

$$(\alpha - \alpha')\frac{2u}{u^2 + v^2 + 1} + (\beta - \beta')\frac{2v}{u^2 + v^2 + 1} + (\gamma - \gamma')\frac{u^2 + v^2 - 1}{u^2 + v^2 + 1} = 0.$$

This simplifies to

$$\left(u + \frac{\alpha - \alpha'}{\gamma - \gamma'} \right)^2 + \left(v + \frac{\beta - \beta'}{\gamma - \gamma'} \right)^2 = 1 + \left(\frac{\alpha - \alpha'}{\gamma - \gamma'} \right)^2 + \left(\frac{\beta - \beta'}{\gamma - \gamma'} \right)^2, \tag{2}$$

which is a circle with centre $\left(-\frac{\alpha - \alpha'}{\gamma - \gamma'}, -\frac{\beta - \beta'}{\gamma - \gamma'} \right)$ and radius

$$\rho = \sqrt{1 + \left(\frac{\alpha - \alpha'}{\gamma - \gamma'} \right)^2 \left(\frac{\beta - \beta'}{\gamma - \gamma'} \right)^2} = \sqrt{2(1 - \alpha\alpha' - \beta\beta' - \gamma\gamma')}/(\gamma - \gamma').$$

The latter form of ρ comes from the fact that $\alpha^2 + \beta^2 + \gamma^2 = \alpha'^2 + \beta'^2 + \gamma'^2 = 1$ because P, $P' \in \mathbb{S}^2$. Of course, these formulas are valid only when $\gamma \neq \gamma'$. When $\gamma = \gamma'$, the image "circle" is the line

$$(\alpha - \alpha')u + (\beta - \beta')v = 0. \tag{3}$$

We shall now show that the stereographic images of P, P' are inverses in the circle (2) or (3).

By Section 3.3, the stereographic images of P, P' are

$$\left(\frac{\alpha}{1 - \gamma}, \frac{\beta}{1 - \gamma} \right), \left(\frac{\alpha'}{1 - \gamma'}, \frac{\beta'}{1 - \gamma'} \right);$$

hence their displacement vectors from the center of (2) are, respectively,

$$\left(\frac{\alpha - \alpha'}{\gamma - \gamma'} + \frac{\alpha}{1 - \gamma}, \frac{\beta - \beta'}{\gamma - \gamma'} + \frac{\beta}{1 - \gamma} \right)$$

$$= \frac{1}{(\gamma - \gamma')(1 - \gamma)}(\alpha - \alpha' + \alpha'\gamma - \alpha\gamma', \beta - \beta' + \beta'\gamma - \beta\gamma')$$

and

$$\left(\frac{\alpha - \alpha'}{\gamma - \gamma'} + \frac{\alpha'}{1 - \gamma'}, \frac{\beta - \beta'}{\gamma - \gamma'} + \frac{\beta'}{1 - \gamma'}\right)$$

$$= \frac{1}{(\gamma - \gamma')(1 - \gamma')}(\alpha - \alpha' + \alpha'\gamma - \alpha\gamma', \beta - \beta' + \beta'\gamma - \beta\gamma').$$

Since these vectors are proportional, the images of P, P' lie on ray through the center of the circle (2). Also, the product of the lengths of these vectors is

$$\frac{1}{(\gamma - \gamma')^2(1 - \gamma)(1 - \gamma')}[(\alpha - \alpha' + \alpha'\gamma - \alpha\gamma')^2 + (\beta - \beta' + \beta'\gamma - \beta\gamma')^2]$$

$$= \frac{1}{(\gamma - \gamma')^2(1 - \gamma)(1 - \gamma')}\{[\alpha(1 - \gamma') - \alpha'(1 - \gamma)]^2$$

$$+ [\beta(1 - \gamma') - \beta'(1 - \gamma)]^2\}$$

$$= \frac{1}{(\gamma - \gamma')^2}\left((\alpha^2 + \beta^2)\frac{1 - \gamma'}{1 - \gamma} + (\alpha'^2 + \beta'^2)\frac{1 - \gamma}{1 - \gamma'} - 2\alpha\alpha' - 2\beta\beta'\right)$$

$$= \frac{1}{(\gamma - \gamma')^2}\left((1 - \gamma^2)\frac{1 - \gamma'}{1 - \gamma} + (1 - \gamma'^2)\frac{1 - \gamma}{1 - \gamma'} - 2\alpha\alpha' - 2\beta\beta'\right)$$

since $\alpha^2 + \beta^2 + \gamma^2 = \alpha'^2 + \beta'^2 + \gamma'^2 = 1$

$$= (2 - 2\alpha\alpha' - 2\beta\beta' - 2\gamma\gamma')/(\gamma - \gamma')^2 = \rho^2 \quad \text{as required.}$$

Finally, if $\gamma = \gamma'$, then the stereographic images of P, P' are

$$\left(\frac{\alpha}{1 - \gamma}, \frac{\beta}{1 - \gamma}\right), \quad \left(\frac{\alpha'}{1 - \gamma}, \frac{\beta'}{1 - \gamma}\right).$$

The equidistant line of these two points is easily seen to be (3), and hence they are exchanged by reflection in (3).   □

Remark. It should be mentioned that the case $\gamma = 1$ (or $\gamma' = 1$) is ignored in the above proof because this means P (or P') is the north pole, hence not in the domain of stereographic projection. What is noteworthy, however, is that $\gamma = 1$ implies $\alpha = \beta = 0$, in which case the stereographic image $\left(\frac{\alpha'}{1-\gamma'}, \frac{\beta'}{1-\gamma'}\right)$ of P' is the center of the circle of inversion, hence lacking an inverse. Thus, stereographic projection and inversion conspire, as it were, to send P to infinity. By adding a point ∞ to the euclidean plane $\mathbb{C}$ we can simultaneously extend stereographic projection to all of $\mathbb{S}^2$ (by making ∞ the image of the north pole) and inversion to all of $\mathbb{C} \cup \{\infty\}$ (by making ∞ inverse to the center of the circle of inversion), and ensure that the theorem remains true under this extended interpretation.

To interpret inversion as a complex function we let $(u, v) = u + iv = w$. Then the simplest case is where the circle C of inversion is the unit circle $|w| = 1$. The inverse of $w = \rho e^{i\theta}$, where $\rho, \theta \in \mathbb{R}$, is obviously $\rho^{-1}e^{i\theta} = 1/\bar{w}$. Hence, inversion in C is the function $I(w) = 1/\bar{w}$. This function enables us to give an easy derivation of the following:

Geometric Properties of Inversion. *Inversion in a circle*

(i) *maps circles to circles,*

(ii) *preserves the magnitude of angles (but reverses their sign).*

Proof. It is clear from the definition of inversion that inverses with respect to a circle C are independent of scale and the choice of origin. Hence, there is no loss of generality in assuming C to be the unit circle, so that inversion is the function $I(w) = 1/\bar{w}$ [unless, of course, the "circle" is a line, in which case inversion is reflection and properties (i) and (ii) are obvious].

(i) The circle $C_{d,\rho}$ with center $d \in \mathbb{C}$ and radius $\rho \in \mathbb{R}$ has equation

$$\rho^2 = |w - d|^2 = (w - d)(\bar{w} - \bar{d}) = w\bar{w} - \bar{d}w - d\bar{w} + d\bar{d},$$

which can be rewritten

$$w\bar{w} - \bar{d}w - d\bar{w} + \sigma = 0,$$

where $d \in \mathbb{C}$ and $\sigma = d\bar{d} - \rho^2 \in \mathbb{R}$ are arbitrary. The function I (which is self-inverse) maps this circle to

$$1/\bar{w}w - \bar{d}/\bar{w} - d/w + \sigma = 0,$$

i.e.,

$$1 - \bar{d}w - d\bar{w} + \sigma w\bar{w} = 0,$$

which is still a circle (a line when $\sigma = 0$).

(ii) I is the product of $w \mapsto 1/w$ and $w \mapsto \bar{w}$. The latter function obviously preserves the magnitude of angles, but reverses their sign; hence it suffices to show that $w \mapsto 1/w$ preserves angles. This happens on general grounds of differentiability, but it can also be seen directly from the calculation

$$\frac{1}{w + \delta e^{i\theta}} - \frac{1}{w} = \frac{-\delta e^{i\theta}}{w(w + \delta e^{i\theta})} \simeq \frac{-\delta e^{i\theta}}{w^2}$$

when δ is small. Thus, when $w + \delta e^{i\theta}$ approaches w from direction θ, $1/(w + \delta e^{i\theta})$ approaches $1/w$ from direction $\theta + \pi + \arg(w^{-2})$, which is $\theta +$ constant. It follows that the angle between different directions is preserved. □

Corollary. *Stereographic projection maps circles to circles and preserves the magnitude of angles.*

Proof. If C is any circle on $\mathbb{S}^2$ with center at the south pole S, then its stereographic image C' is obviously also a circle. Sending S to any other point $P \in \mathbb{S}^2$ by reflection in the equidistant plane of S and P sends C to an arbitrary circle D with center P, and at the same time sends its stereographic image C' to the stereographic image D' of D by an inversion. Hence, D' is also a circle.

Likewise, any angle at the south pole is obviously preserved by stereographic projection. Sending the angle to P by reflection maps its stereographic image by an inversion, and hence preserves its magnitude.   □

Exercises

3.4.1. Show that reflection is a limiting case of inversion.

3.4.2. Show that the stereographic images of antipodal points are of the form $w, -1/\bar{w}$.

3.5 Reflections and Rotations as Complex Functions

If we introduce the notation

$$\lambda = \alpha - \alpha', \quad \mu = \beta - \beta', \quad \nu = \gamma - \gamma',$$

then the content of the theorem of Section 3.4 can be restated as follows: reflection of $\mathbb{S}^2$ in the plane

$$\lambda x + \mu y + \nu z = 0 \tag{1}$$

induces inversion of $\mathbb{C}$ in the circle $C_{d,\rho}$ with

$$\text{center } d = -\lambda/\nu - i\mu/\nu \tag{2}$$

and

$$\text{radius } \rho = \sqrt{1 + (\lambda/\nu)^2 + (\mu/\nu)^2} = (1/\nu)\sqrt{\lambda^2 + \mu^2 + \nu^2}.$$

Thus, if we normalize the coefficients of (1) so that $\lambda^2 + \mu^2 + \nu^2 = 1$, we have

$$\text{radius } \rho = 1/\nu. \tag{3}$$

Proposition. *The map of the w-plane induced by reflection of $\mathbb{S}^2$ in the plane $\lambda x + \mu y + \nu z = 0$, where $\lambda^2 + \mu^2 + \nu^2 = 1$, is*

$$\bar{g}(w) = \frac{(\lambda + i\mu)\bar{w} - \nu}{-\nu\bar{w} - (\lambda - i\mu)}.$$

Proof. It is immediate from the definition of inversion that inverses with respect to a circle C centered on O are preserved by the dilatation $d_\rho(w) = \rho w$ for any nonzero $\rho \in \mathbb{R}$, and the translation $t_d(w) = d + w$ for any $d \in \mathbb{C}$. Since $C_{d,\rho}$ is obtained from the unit circle by $t_d d_\rho$, inversion in $C_{d,\rho}$ is the conjugate $t_d d_\rho I d_\rho^{-1} t_d^{-1}$ of inversion I in the unit circle by $t_d d_\rho$.

Since $I(w) = 1/\bar{w}$ by Section 3.4, it follows that $t_d d_\rho I d_\rho^{-1} t_d^{-1}$ is the function

$$\bar{g}(w) = d + \rho \left(\frac{1}{\rho^{-1}(-d+w)} \right) = \frac{d\bar{w} + \rho^2 - d\bar{d}}{\bar{w} - \bar{d}}.$$

By (2) and (3) we have $d = -\lambda/\nu - i\mu/\nu$, $\rho = 1/\nu$; hence

$$
\begin{aligned}
\bar{g}(w) &= \frac{(\lambda + i\mu)\bar{w} - (1 - \lambda^2 - \mu^2)/\nu}{-\nu\bar{w} - (\lambda - i\mu)} \\
&= \frac{(\lambda + i\mu)\bar{w} - \nu^2/\nu}{-\nu\bar{w} - (\lambda - i\mu)} \quad \text{since } \lambda^2 + \mu^2 + \nu^2 = 1 \\
&= \frac{(\lambda + i\mu)\bar{w} - \nu}{-\nu\bar{w} - (\lambda - i\mu)}. \qquad \square
\end{aligned}
$$

The two real parameters λ, μ, which are arbitrary except for the constraint $\lambda^2 + \mu^2 + \nu^2 = 1$, can be replaced by the complex parameter

$$l = \lambda + i\mu,$$

which is arbitrary except for the constraint $|l|^2 + \nu^2 = 1$. This simplifies the calculations somewhat when we compute the product of two reflections, i.e., a rotation.

Theorem (Gauss [c. 1819]). *The maps of the w-plane $\mathbb{C}$ induced by rotations of $\mathbb{S}^2$ are precisely the functions*

$$f(w) = \frac{aw + b}{-\bar{b}w + \bar{a}},$$

where $a, b \in \mathbb{C}$ and $|a|^2 + |b|^2 = 1$.

Proof. A rotation is the product of reflections, which we can take to be

$$\bar{g}_1(w) = \frac{l_1\bar{w} - \nu_1}{-\nu_1\bar{w} - \bar{l}_1}, \quad \bar{g}_2(w) = \frac{l_2\bar{w} - \nu_2}{-\nu_2\bar{w} - \bar{l}_2},$$

where $|l_1|^2 + \nu_1^2 = |l_2|^2 + \nu_2^2 = 1$ by the proposition. Then a few lines of calculation yield

$$f(w) = \bar{g}_1\bar{g}_2(w) = \frac{(l_1\bar{l}_2 + \nu_1\nu_2)w + (-l_1\nu_2 + \nu_1 l_2)}{(\bar{l}_1\nu_2 - \nu_1\bar{l}_2)w + (\bar{l}_1 l_2 + \nu_1\nu_2)}.$$

This is of the form $\frac{aw+b}{-\bar{b}w+\bar{a}}$ with $a = l_1\bar{l}_2 + \nu_1\nu_2$ and $b = -l_1\nu_2 + \nu_1 l_2$. Also

$$
\begin{aligned}
|a|^2 + |b|^2 = a\bar{a} + b\bar{b} &= \det\begin{pmatrix} a & b \\ -\bar{b} & \bar{a} \end{pmatrix} = \det\begin{pmatrix} l_1\bar{l}_2 + \nu_1\nu_2 & -l_1\nu_2 + \nu_1 l_2 \\ \bar{l}_1\nu_2 - \nu_1\bar{l}_2 & \bar{l}_1 l_2 + \nu_1\nu_2 \end{pmatrix} \\
&= \det\begin{pmatrix} l_1 & -\nu_1 \\ -\nu_1 & -\bar{l}_1 \end{pmatrix}\begin{pmatrix} \bar{l}_2 & -\nu_2 \\ -\nu_2 & -l_2 \end{pmatrix} \\
&= \det\begin{pmatrix} l_1 & -\nu_1 \\ -\nu_1 & -\bar{l}_1 \end{pmatrix}\det\begin{pmatrix} \bar{l}_2 & -\nu_2 \\ -\nu_2 & -l_2 \end{pmatrix} \\
&= (-|l_1|^2 - \nu_1^2)(-|l_2|^2 - \nu_2^2) = (-1)(-1) = 1.
\end{aligned}
$$

Hence, any rotation is of the required form.

Conversely, we can get arbitrary complex coefficients

$$ a = l_1\bar{l}_2 + \nu_1\nu_2, \quad b = -l_1\nu_2 + \nu_1 l_2, $$

such that $|a|^2 + |b|^2 = 1$ by suitable choice of l_1, l_2, ν_1, ν_2. For example, take $\nu_2 = 0$, which implies $|l_2| = 1$ so $l_2 = e^{i\phi}$ for some $\phi \in \mathbb{R}$. Then

$$ a = l_1 e^{-i\phi}, \quad b = \nu_1 e^{i\phi}. $$

Now write $l_1 = \lambda_1 e^{i\theta}$, so $|l_1|^2 + \nu_1^2 = \lambda_1^2 + \nu_1^2 = 1$ and

$$ a = \lambda_1 e^{i(\theta-\phi)}, \quad b = \nu_1 e^{i\phi}. $$

Since λ_1, ν_1 are arbitrary reals such that $\lambda_1^2 + \nu_1^2 = 1$, and $\theta - \phi$, ϕ are an arbitrary pair of angles, this shows a, b to be arbitrary complex numbers such that $|a|^2 + |b|^2 = 1$. □

Corollary. *The maps of the w-plane $\mathbb{C}$ induced by orientation-reversing isometries of $\mathbb{S}^2$ are precisely the functions*

$$ \bar{f}(w) = \frac{a\bar{w} + b}{-\bar{b}\bar{w} + \bar{a}}, $$

where $a, b \in \mathbb{C}$ and $|a|^2 + |b|^2 = 1$.

Proof. By the remarks following the theorem of Section 3.2, any orientation-reversing isometry of $\mathbb{S}^2$ is of the form $r\bar{r}$, where r is a rotation and $\bar{r}$ is a fixed reflection which can be chosen arbitrarily. A convenient $\bar{r}$ to choose is reflection in the plane $y = 0$, which induces the map $\bar{g}(w) = \bar{w}$ of the w-plane. It then follows from the theorem that the map induced by $r\bar{r}$ is

$$ f\bar{g}(w) = \frac{a\bar{w} + b}{-\bar{b}\bar{w} + \bar{a}}. \qquad \square $$

Remarks. The appearance of 2×2 matrices in the above proof is not a fluke, nor a clever shortcut of the author. Any function of the form

$$ f(w) = \frac{aw + b}{cw + d} $$

(called *linear fractional*) behaves in much the same way as the matrix $\begin{pmatrix} a & b \\ c & d \end{pmatrix}$. In particular, if

$$f_1(w) = \frac{a_1 w + b_1}{c_1 w + d_1}, \quad f_2(w) = \frac{a_2 w + b_2}{c_2 w + d_2},$$

then

$$f_1 f_2(w) = \frac{aw + b}{cw + d}$$

where

$$\begin{pmatrix} a & b \\ c & d \end{pmatrix} = \begin{pmatrix} a_1 & b_1 \\ c_1 & d_1 \end{pmatrix} \begin{pmatrix} a_2 & b_2 \\ c_2 & d_2 \end{pmatrix}.$$

Of course, many matrices correspond to the same fraction because numerator and denominator can be multiplied by any nonzero constant. The matrices obtained in the theorem, with $|a|^2 + |b|^2 = 1$, are the most convenient to use because

$$|a|^2 + |b|^2 = a\bar{a} + b\bar{b} = \det \begin{pmatrix} a & b \\ -\bar{b} & \bar{a} \end{pmatrix}.$$

Since the determinant of a product is the product of the determinants, this guarantees that the product of two such matrices will be another of the same type.

Exercises

3.5.1. Rewrite the fraction $\frac{(\lambda + i\mu)\bar{w} - \nu}{-\nu\bar{w} - (\lambda - i\mu)}$ in the form $\frac{a\bar{w} + b}{-\bar{b}\bar{w} + \bar{a}}$ given by the corollary.

3.5.2. Check that if

$$f_1(w) = \frac{a_1 w + b_1}{c_1 w + d_1}, \quad f_2(w) = \frac{a_2 w + b_2}{c_2 w + d_2},$$

then

$$f_1 f_2(w) = \frac{aw + b}{cw + d}$$

where

$$\begin{pmatrix} a & b \\ c & d \end{pmatrix} = \begin{pmatrix} a_1 & b_1 \\ c_1 & d_1 \end{pmatrix} \begin{pmatrix} a_2 & b_2 \\ c_2 & d_2 \end{pmatrix}.$$

3.5.3. Show that a rotation of $\mathbb{S}^2$ about the y-axis induces a map of the w-plane of the form $f(w) = \frac{\nu w - \lambda}{\lambda w + \nu}$, where $\lambda^2 + \nu^2 = 1$ (e.g., by expressing the rotation as the product of reflections in the planes $\lambda x + \nu z = 0$ and $z = 0$).

3.5.4. Using the map $r_\theta(w) = e^{i\theta}w$ induced by rotation of $\mathbb{S}^2$ about the z-axis through θ, compute an expression for

(rotation by ϕ about z-axis)(rotation about y-axis)
(rotation by θ about z-axis).

3.5.5. Show that the expression obtained in Exercise 3.5.4 represents an arbitrary rotation of $\mathbb{S}^2$.

3.6   The Antipodal Map and the Elliptic Plane

Are there any surfaces "locally like" $\mathbb{S}^2$ which are not themselves spheres? We can tackle this question by the approach used in Chapter 2 to find surfaces "locally like" the euclidean plane. We define a *spherical surface S* to be one in which each point has an ϵ-neighborhood isometric to a disc of $\mathbb{S}^2$, show that a complete, connected spherical S can be covered by $\mathbb{S}^2$, and finally express S as $\mathbb{S}^2/\Gamma$, where Γ is a discontinuous, fixed point free group of isometries of $\mathbb{S}^2$. The only point where the argument of Chapter 2 needs some modification is in the construction of the covering $\mathbb{S}^2 \to S$. As before, we obtain the covering as a "pencil map" p defined by filling $\mathbb{S}^2$ with a pencil of rays emanating from some point $O \in \mathbb{S}^2$. The problem is that these rays (great circles) meet again at distance π from O (at the antipodal point O' of O), whereas it is conceivable that their p-images on S tend to *different* points at distance π from $p(O)$. This possibility is fortunately ruled out by the local isometry property—points near O' must be mapped to points which are close together on S.

Thus, the problem of finding complete, connected spherical surfaces reduces to the problem of finding the possible groups Γ. Since Γ must be fixed point free, this immediately rules out groups containing nontrivial rotations, as these always have fixed points. Since the square of any orientation-reversing isometry is a rotation, by the remarks following the theorem of Section 3.2, the only possibility for Γ is $\{1 = \text{identity}, f\}$, where f is a fixed point free, orientation-reversing isometry such that $f^2 = 1$.

Theorem. *The only fixed point free, orientation-reversing isometry f such that $f^2 = 1$ is the antipodal map*

$$\bar{m} : (x, y, z) \mapsto (-x, -y, -z).$$

Proof. Since $\bar{m}\{\text{rotations}\} = \{\text{orientation-reversing isometries}\}$ by the remarks following the theorem of Section 3.2, any orientation-reversing isometry is of the form $\bar{m}r$, where r is a rotation. We choose axes so that the z-axis is the axis of r (Figure 3.6).

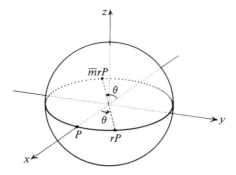

FIGURE 3.6.

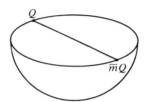

FIGURE 3.7.

If P is on the "equator" of $\mathbb{S}^2$ (intersection of $\mathbb{S}^2$ with the plane $z = 0$), and if r is a rotation through θ, we see that $\bar{m}rP$ lies on the equator at angle $\theta + \pi$ from P, and $(\bar{m}r)^2P$ at angle 2θ from P. Hence, $(\bar{m}r)^2 = 1$ implies $\theta = 0$ or π. But if $\theta = \pi$, then $\bar{m}rP = P$; hence $\bar{m}r$ has fixed points. Thus the only possibility is $\theta = 0$, in which case r is the identity, so $\bar{m}r$ is simply $\bar{m}$.   □

Corollary. *The only complete connected spherical surface other than $\mathbb{S}^2$ is $\mathbb{S}^2/\{1, \bar{m}\}$, where $\bar{m}$ is the antipodal map.*

The surface $\mathbb{S}^2/\{1, \bar{m}\}$ is called the *elliptic plane*. Its points are the $\{1, \bar{m}\}$-orbits, i.e., the antipodal point pairs of $\mathbb{S}^2$. Intuitively speaking, the elliptic plane is obtained from a hemisphere by identifying all antipodal points Q, $\bar{m}Q$ on the equator (Figure 3.7). This surface cannot be constructed in $\mathbb{R}^3$, even if lengths are distorted, without intersecting itself. In this respect it is similar to the twisted cylinder and the Klein bottle (Sections 2.3 and 2.4). In fact, like them, the elliptic plane contains a Möbius band (Exercise 3.6.3), though, of course, not isometrically—only topologically.

Like the sphere, the elliptic plane can be mapped onto the euclidean plane in a way which illuminates its geometric properties. The appropriate

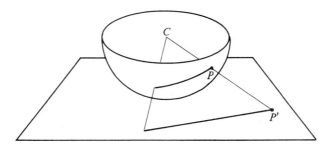

FIGURE 3.8.

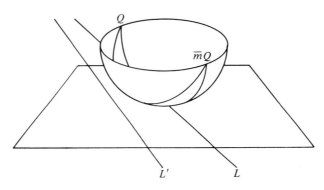

FIGURE 3.9.

map in this case is *central projection* (Fig. 3.8), sending each P in the hemisphere to the point P' where the line CP from the center of the sphere meets the horizontal tangent plane. Central projection maps the "lines" of the elliptic plane (great semicircles) to lines of the plane, though failing to provide images for points of the equator.

Once again it is desirable to add "infinity" to the plane to represent the points which lack images. Each antipodal pair Q, $\bar{m}Q$ on the equator corresponds to the pair of "ends at infinity" of a line L through the south pole (Figure 3.9). Since Q, $\bar{m}Q$ are the same point on the elliptic plane, we have to agree that the two ends of L are the *one* "point at infinity", and this "point at infinity" has to be shared by all lines L' parallel to L, as these are the images of the great semicircles through Q, $\bar{m}Q$.

Adding a "point at infinity" to the plane for each family of parallels is also what one does in projective geometry—one wants parallels to "meet at infinity"—hence, the elliptic plane is also called the *projective plane*. (However, the class of mappings natural to the projective plane—the *collineations* or mappings which send lines to lines—is larger than the class of isometries of the elliptic plane. Whereas each isometry of the elliptic plane induces a collineation of the plane under central projection, the converse is not true. See Exercise 3.6.4.)

Exercises

3.6.1. Imitate the argument of the theorem of Section 1.5 to show that each orientation-reversing isometry of $\mathbb{S}^2$ is a "glide reflection", i.e., the product of a reflection in a line L^* with a rotation which maps L^* onto itself.

3.6.2. Show that the antipodal map is the only glide reflection which maps more than one line onto itself.

3.6.3. Find a topological Möbius band in the elliptic plane.

3.6.4. Show that the dilatation: $(x, y) \mapsto (rx, ry)$, for $1 \neq r \in \mathbb{R}$, is a collineation of the plane not induced by an isometry of the elliptic plane [where the elliptic plane is mapped onto the (x, y)-plane by central projection].

3.7 Remarks on Groups, Spheres, and Projective Spaces

To see the group of rotations of $\mathbb{S}^2$ in some perspective, one needs to be aware of "spheres" in all euclidean spaces $\mathbb{R}^n$ with $n \geq 2$. It is reasonable to call the unit circle $x^2 + y^2 = 1$ in the xy-plane the 1-*sphere* $\mathbb{S}^1$ because it is the analogue in $\mathbb{R}^2$ of $\mathbb{S}^2$ in $\mathbb{R}^3$. In this terminology, the ordinary sphere $\mathbb{S}^2$ is called the 2-*sphere*, the analogous figure $x^2 + y^2 + z^2 + t^2 = 1$ in the (x, y, z, t)-space $\mathbb{R}^4$ is called the 3-*sphere*, and so on.

The group of rotations of the n-sphere $\mathbb{S}^n$ is called $\mathrm{SO}(n + 1)$, where SO stands for "special orthogonal" and we have $n + 1$, rather than n, because the rotations of $\mathbb{S}^n$ are viewed as orientation-preserving isometries of $\mathbb{R}^{n+1}$ which leave O fixed. Thus, $\mathrm{SO}(2)$ is the group of rotations of the circle $\mathbb{S}^1$, and $\mathrm{SO}(2)$ *itself* can be viewed as an $\mathbb{S}^1$ because each rotation of $\mathbb{S}^1$ corresponds to an angle, and hence to a point of $\mathbb{S}^1$.

This raises the question: can $\mathrm{SO}(3)$, the group of rotations of $\mathbb{S}^2$, also be viewed as a geometric space, and if so, what is it? To answer this question we use the representation of rotations by fractions of the form $\frac{aw+b}{-bw+\bar{a}}$, where $|a|^2 + |b|^2 = 1$. Writing $a = \alpha_1 + i\alpha_2$ and $b = \beta_1 + i\beta_2$, we see that each rotation is determined by a point $(\alpha_1, \alpha_2, \beta_1, \beta_2) \in \mathbb{R}^4$ which satisfies the condition $\alpha_1^2 + \alpha_2^2 + \beta_1^2 + \beta_2^2 = 1$. These points form a 3-sphere!

However, we have to remember that, even under the condition $|a|^2 + |b|^2 = 1$, there are *two* formally different fractions which represent the same function, namely,

$$\frac{aw + b}{-\bar{b}w + \bar{a}} \quad \text{and} \quad \frac{-aw - b}{\bar{b}w - \bar{a}}.$$

The latter fraction corresponds to the point $(-\alpha_1, -\alpha_2, -\beta_1, -\beta_2) \in \mathbb{S}^3$, "antipodal" to $(\alpha_1, \alpha_2, \beta_1, \beta_2)$. Thus, the rotations of $\mathbb{S}^2$ actually correspond to the *antipodal point pairs* in an $\mathbb{S}^3$. Just as we call the space of antipodal point pairs on $\mathbb{S}^2$ the elliptic plane or the projective plane [denoted by $P_2(\mathbb{R})$], the space of antipodal point pairs in $\mathbb{S}^3$ is called *elliptic space* or *projective space*, and denoted by $P_3(\mathbb{R})$.

Thus, SO(2) can be viewed as $\mathbb{S}^1$, and SO(3) can be viewed as $P_3(\mathbb{R})$. Looking at these results from the opposite end, we have found *group structures* on $\mathbb{S}^1$ and $P_3(\mathbb{R})$; that is, rules for "multiplying points" in these spaces, under which the spaces become groups. The rule for multiplying points $\theta, \phi \in \mathbb{S}^1$ is simply $\theta, \phi \mapsto \theta + \phi$, whereas the rule for multiplying $(\pm\alpha_1, \pm\alpha_2, \pm\beta_1, \pm\beta_2)$, $(\pm\alpha_1', \pm\alpha_2', \pm\beta_1', \pm\beta_2') \in P_3(\mathbb{R})$ is to form the corresponding fractions $\frac{aw+b}{-\bar{b}w-\bar{a}}$, $\frac{a'w+b'}{-\bar{b}w+\bar{a}}$, substitute the second in the first to get $\frac{a''w+b''}{-\bar{b}''w+\bar{a}''}$ say, then extract $(\pm\alpha_1'', \pm\alpha_2'', \pm\beta_1'', \pm\beta_2'')$ from $a'' = \alpha_1'' + i\alpha_2''$ and $b'' = \beta_1'' + i\beta_2''$.

While there are many quite uninteresting ways to give group structure to arbitrary sets of points, the group structures just found on $\mathbb{S}^1$ and $P_3(\mathbb{R})$ are interesting because the functions which "multiply points" are continuous. It is relatively rare for a space to possess such a group structure, and, in fact, $\mathbb{S}^1$ and $\mathbb{S}^3$ are the *only* spheres which possess continuous group structures.

We have not yet exhibited a continuous group structure on $\mathbb{S}^3$ but, not surprisingly, it is related to the one on $P_3(\mathbb{R})$. Instead of considering the fractions $\frac{aw+b}{-\bar{w}+\bar{a}}$ with $|a|^2 + |b|^2 = 1$, we consider the matrices $\begin{pmatrix} a & b \\ -\bar{b} & \bar{a} \end{pmatrix}$. As shown in Exercise 3.5.2, substitution in the fractions corresponds to multiplication of the matrices. However, the matrices correspond one-to-one, not two-to-one, with the points $(\alpha_1, \alpha_2, \beta_1, \beta_2) \in \mathbb{S}^3$. Thus, we get a continuous group structure on $\mathbb{S}^3$ by letting $(\alpha_1, \alpha_2, \beta_1, \beta_2)$ correspond to the matrix $\begin{pmatrix} \alpha_1 + i\alpha_2 & \beta_1 + i\beta_2 \\ -\beta_1 + i\beta_2 & \alpha_1 - i\alpha_2 \end{pmatrix}$, and "multiplying points" by multiplying the corresponding matrices.

The matrix $\begin{pmatrix} \alpha_1 + i\alpha_2 & \beta_1 + i\beta_2 \\ -\beta_1 + i\beta_2 & \alpha_1 - i\alpha_2 \end{pmatrix}$ can be written as $\alpha_1 1 + \alpha_2 i + \beta_1 j + \beta_2 k$, where

$$1 = \begin{pmatrix} 1 & 0 \\ 0 & 1 \end{pmatrix}, \quad i = \begin{pmatrix} i & 0 \\ 0 & -i \end{pmatrix}, \quad j = \begin{pmatrix} 0 & 1 \\ -1 & 0 \end{pmatrix}, \quad k = \begin{pmatrix} 0 & i \\ i & 0 \end{pmatrix}.$$

The matrices $1, i, j, k$ are called the *quaternion units*, following Hamilton, who introduced abstract quantities i, j, k satisfying the laws $i^2 = j^2 = k^2 = ijk = -1$ to make an algebraic analysis of rotations in $\mathbb{R}^3$. Hamilton gave the name *quaternions* to quantities of the form $\alpha + \beta i + \gamma j + \delta k$, where $\alpha, \beta, \gamma, \delta \in \mathbb{R}$. His laws for multiplying i, j, k determine a multiplication of quaternions which amounts to the same thing as matrix multiplication with i, j, k taken to be the matrices above. Thus, if the points of $\mathbb{S}^3$ are taken to be the quaternions $\alpha + \beta i + \gamma j + \delta k$ with *norm* $\alpha^2 + \beta^2 + \gamma^2 + \delta^2$ equal to 1, then the group operation on $\mathbb{S}^3$ is simply quaternion multiplication.

Exercises

3.7.1. Assuming the obvious definition of $P_1(\mathbb{R})$, show that $P_1(\mathbb{R})$ is homeomorphic to $\mathbb{S}^1$, and that the isometries of $P_1(\mathbb{R})$ form a $P_1(\mathbb{R})$.

3.7.2. Show that a reflection of $\mathbb{S}^2$ induces a rotation of $P_2(\mathbb{R})$ through π.

3.7.3. Show that the isometries of $P_2(\mathbb{R})$ form a $P_3(\mathbb{R})$.

3.8 The Area of a Triangle

Until now we have taken it for granted that the sphere is not locally isometric to the plane. This belief can be confirmed by finding a local geometric property of the sphere $\mathbb{S}^2$ which does not hold in any euclidean disc. The *lines* of $\mathbb{S}^2$ can be defined, as can the lines of the euclidean plane, as the fixed point sets of reflections (which can, in turn, be defined, for both surfaces, as the nontrivial isometries with more than two fixed points). Thus, the lines of $\mathbb{S}^2$ are precisely the great circles. The *angle* between two great circles C_1, C_2 is the angle between the planes Π_1, Π_2 containing them, which is also the angle between the tangents to C_1, C_2 at their point of intersection (Figure 3.10). We know (Section 1.6) that the angle sum of any euclidean triangle is π. In contrast, the sphere has the following property:

Theorem (Harriot [1603]). *For any spherical triangle Δ, with angles α, β, γ, the function*

$$\text{excess}(\Delta) = \alpha + \beta + \gamma - \pi$$

is proportional to area(Δ), *and hence nonzero.*

Proof. We assume only the following properties of area:

(i) The area of a sector of angle α between two great circles, as shown in Figure 3.10, is $\alpha/2\pi$ of the area of $\mathbb{S}^2$.

(ii) Area is invariant under isometries.

(iii) Area is additive.

Now let $\Delta_{\alpha\beta\gamma}$ be any spherical triangle with angles α, β, γ. By prolonging the sides of $\Delta_{\alpha\beta\gamma}$ to great circles, we get a partition of $\mathbb{S}^2$ into eight triangles, as shown in Figure 3.11.

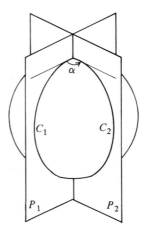

FIGURE 3.10.

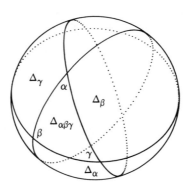

FIGURE 3.11.

The four shown on the "front" of the sphere—$\Delta_{\alpha\beta\gamma}$, Δ_α, Δ_β, Δ_γ—are paired with isometric antipodal triangles on the "back." Since $\Delta_{\alpha\beta\gamma} \cup \Delta_\alpha$ is a sector of angle α, we have

$$\text{area}(\Delta_{\alpha\beta\gamma}) + \text{area}(\Delta_\alpha) = \alpha/2\pi \ \text{area}(\mathbb{S}^2) \tag{1}$$

by (i). Similarly we get

$$\text{area}(\Delta_{\alpha\beta\gamma}) + \text{area}(\Delta_\beta) = \beta/2\pi \ \text{area}(\mathbb{S}^2), \tag{2}$$

$$\text{area}(\Delta_{\alpha\beta\gamma}) + \text{area}(\Delta_\gamma) = \gamma/2\pi \ \text{area}(\mathbb{S}^2). \tag{3}$$

Adding equations (1), (2), (3) gives

$$3\text{area}(\Delta_{\alpha\beta\gamma}) + \text{area}(\Delta_\alpha) + \text{area}(\Delta_\beta) + \text{area}(\Delta_\gamma) = \frac{\alpha + \beta + \gamma}{2\pi} \ \text{area}(\mathbb{S}^2). \tag{4}$$

On the other hand, adding the areas of $\Delta_{\alpha\beta\gamma}$, Δ_α, Δ_β, Δ_γ to those of their antipodal triangles gives $\text{area}(\mathbb{S}^2)$; hence, by (ii),

$$\text{area}(\Delta_{\alpha\beta\gamma}) + \text{area}(\Delta_\alpha) + \text{area}(\Delta_\beta) + \text{area}(\Delta_\gamma) = \tfrac{1}{2} \ \text{area}(\mathbb{S}^2). \tag{5}$$

Subtracting (5) from (4) gives

$$2\,\text{area}(\Delta_{\alpha\beta\gamma}) = \frac{\alpha + \beta + \gamma - \pi}{2\pi} \ \text{area}(\mathbb{S}^2);$$

hence

$$\text{area}(\Delta_{\alpha\beta\gamma}) = \text{excess}(\Delta_{\alpha\beta\gamma}) \frac{\text{area}(\mathbb{S}^2)}{4\pi},$$

which proves the theorem. (In fact, assuming the result from calculus that $\text{area}(\mathbb{S}^2) = 4\pi$, we have area = excess.) $\square$

Exercises

3.8.1. Define excess(Σ) for a spherical polygon Σ, and hence find area(Σ).

3.8.2. Show that two spherical triangles with the same angles are isometric.

3.8.3. Show that the distance between points $P, Q \in \mathbb{S}^2$ is the angle between the perpendiculars to the great circle segment PQ at P and Q.

3.9   The Regular Polyhedra

The regular polyhedra are five figures from the classical geometry of $\mathbb{R}^3$ illustrated in Figure 3.12—the tetrahedron, cube, octahedron, dodecahedron, and icosahedron. The faces of the tetrahedron, octahedron and icosahedron are equilateral triangles, those of the cube are squares, and those of

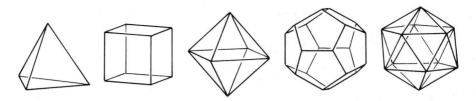

FIGURE 3.12.

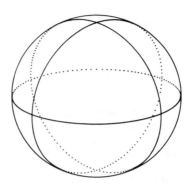

FIGURE 3.13.

the dodecahedron are regular pentagons. It is easy to show that these are the only possible convex polyhedra whose faces are all regular polygons of the same type, because at least three faces must meet at each vertex and, hence, the polygons must have angles $< 2\pi/3$.

Corresponding to each regular polyhedron $\mathcal{P}$ we get a *regular tessellation* of $\mathbb{S}^2$ by placing $\mathcal{P}$ so that its center is at O and projecting the edges of $\mathcal{P}$ from O onto $\mathbb{S}^2$. Each regular polygonal face of $\mathcal{P}$ projects to a regular spherical polygon on $\mathbb{S}^2$. For example, the octahedron projects to the tessellation of $\mathbb{S}^2$ by the eight "octants"—equilateral triangles with angles of $\pi/2$ (Figure 3.13).

Each regular polyhedron $\mathcal{P}$ has a *symmetry group* which is a finite group of rotations of $\mathbb{S}^2$. If we imagine a solid $\mathcal{P}$ occupying a $\mathcal{P}$-shaped "hole" in $\mathbb{R}^3$, then the rotations in the symmetry group are those which turn $\mathcal{P}$ to a position which fits the hole. There are only three such symmetry groups, called the *polyhedral groups*, because the octahedron and cube have the same group, as do the icosahedron and dodecahedron (see Exercise 3.9.1). We shall study these groups further in Chapter 7 when we look more generally at groups associated with regular tessellations.

In addition to the polyhedral groups, there are two infinite families of finite groups of rotations of $\mathbb{S}^2$. The first consists of *cyclic groups* C_n, each generated by a rotation through $2\pi/n$. The second consists of *dihedral groups* D_n. D_n can be regarded as the symmetry group of a degenerate polyhedron—the "dihedron"—with two regular n-agonal faces. D_n has $2n$ elements, generated by two rotations with perpendicular axes, the first rotation being through $2\pi/n$ and the second (which interchanges the two faces of the "dihedron") through π.

The fact that there are only five regular polyhedra has the (not quite obvious) consequence that the only finite groups of rotations of $\mathbb{S}^2$ are C_n, D_n, and the three polyhedral groups. An even more remarkable consequence was proved by Klein [1876]: any finite group of linear fractional transformations is isomorphic to one of C_n, D_n or the three polyhedral groups. Proofs of these theorems may be found in Jones and Singerman [1987, pp. 42–49].

Exercises

3.9.1. Explain the relationship between the cube and octahedron (and dodecahedron and icosahedron) which implies they have the same symmetry group.

3.9.2. Show that each permutation of the four vertices of a tetrahedron can be realized by an isometry, but only the even permutations can be realized by rotations.

3.9.3. Show that each permutation of the four diagonals of a cube can be realized by a rotation, and that the rotation realizing a given permutation of the diagonals is unique.

3.9.4. Show that C_n and D_n, but no other finite groups, can be realized as groups of isometries of the euclidean plane.

3.10 Discussion

The sphere is topologically different from the plane because it is compact and the plane is not. With the help of stereographic projection we can view the plane topologically as the sphere minus its north pole. Then the process of replacing the north pole is called *one point compactification* of the plane. It is not so much the extra point that makes the difference as the fact (implicit in our interpretation of the plane as the sphere minus a point) that the points "near infinity" in the plane are in the neighborhood of the added point, so that nonconvergent sequences on the plane now converge on the sphere.

What the plane and the sphere have in common is *simple connectivity*, a topological property common to all universal covering spaces (see, e.g., Armstrong [1979, p. 233]): A space S is called simply connected if every closed path in S can be "deformed" within S to a point. We shall formalize

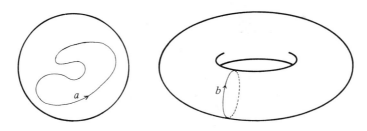

FIGURE 3.14.

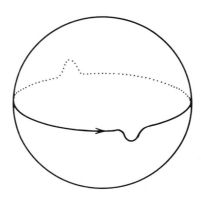

FIGURE 3.15.

the idea of deformation in Chapter 6, but it can be illustrated by the paths a on the sphere and b on the torus shown in Figure 3.14. It is intuitively clear that the path a can be deformed to a point on the sphere, but that b cannot be deformed to a point on the torus. Thus, the torus is *not* simply connected.

Another surface which is not simply connected is the projective plane. If we view the projective plane as the space of antipodal point pairs on the sphere, then it is intuitively plausible that the equatorial path e cannot be contracted to a point. This is because points have to remain antipodal under the deformation, and hence far apart (see, e.g., Figure 3.15).

More surprisingly, the path e^2 which runs *twice* around the equator *can* be deformed to a point. Since e^2 runs around the equator twice, one path "near" to e^2 is represented by two latitude circles e_1, e_2 equidistant from e (Figure 3.16). The union of e_1 and e_2 is indeed a single path on the projective plane because it consists of a continuous sequence of antipodal point pairs. We now easily shrink this path to a point by moving e_1 and e_2 simultaneously to the opposite poles, which, of course, form a single point of the projective plane.

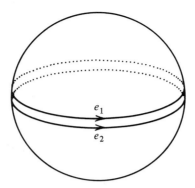

e_1

e_2

FIGURE 3.16.

It will be shown in Chapter 6 that this property is essentially the topological way of saying that the projective plane is the quotient of the sphere by a cyclic group of order 2.

The phenomenon has an interesting counterpart in the 3-sphere $\mathbb{S}^3$ and its space of antipodal point pairs, which we found in Section 3.7 to be the group SO(3) of rotations of $\mathbb{S}^2$. By the same general topological considerations, there should be a path e in SO(3) which is not deformable to a point, whereas e^2 is. This path actually makes a real-life appearance in what is called the "plate trick". I shall describe the trick first, then interpret it by paths in SO(3).

Take a plate filled with soup, and twist it through 2π without spilling any soup (Figure 3.17). You now have a kink in your arm which cannot be removed without twisting the plate, e.g., by reversing the twist just described. But look! You can also remove the kink by a further twist in the *same* direction as the first (Figure 3.18).

This can be explained by viewing the first twist as a closed path e in SO(3), i.e., as a *continuous sequence of rotations* of $\mathbb{S}^2$. Certainly, the arm and plate system moves through a continuous sequence of positions, but how is a given position to be interpreted as a rotation of $\mathbb{S}^2$? The answer is simply that each rotation is determined by the direction of its axis, which we can take to be the direction from the shoulder to the center of the plate, and the amount of turn, which we can take to be the amount of turn of the plate. This interpretation also shows that the sequence of positions in the first twist is a *closed* path e in SO(3) because the plate ends with the same direction and amount of turn it had at the beginning. The second twist e' may not be identical with e, but it is a deformation of it, so the result of the two twists deforms to e^2, which deforms to no movement at all, as Figure 3.18 shows.

(Readers who, like me, did not understand the plate trick the first time it was explained to them may like to read accounts of it in the books of

FIGURE 3.17.

FIGURE 3.18.

Artmann [1988, p. 118], Kostrikin and Manin [1989, p. 169], Kauffman [1987, p. 93], and Montesinos [1987, p. 99]. Kauffman's account has the bonus of relating the plate trick to quaternions.)

Before leaving $\mathbb{S}^3$ it is worth pointing out that this space contains *flat tori*, genuine embedded surfaces isometric to the abstract quotient tori of Chapter 2. The reason is that in $\mathbb{S}^3$, as in $\mathbb{S}^2$, the lines are closed, so a "cylindrical" neighborhood of a line in $\mathbb{S}^3$ is, in fact, a torus. The discovery of concrete flat tori by Clifford [1873] helped appreciably in the development of the abstract surface concept by showing that the abstractions were "real" in a relatively familiar space.

We now return to $\mathbb{C}$ and its compactification $\mathbb{C} \cup \{\infty\}$. The point ∞ used to compactify $\mathbb{C}$ is not only the same topologically as all other points of $\mathbb{C} \cup \{\infty\}$; it is also the same geometrically (because all points of the sphere are alike) *and analytically*. It blends analytically into $\mathbb{C}$ when we define $1/0 = \infty$, as is shown by the fact that the functions $\frac{az+b}{cz+d}$ make sense on $\mathbb{C} \cup \{\infty\}$. (These functions are, in fact, all the conformal mappings of $\mathbb{C} \cup \{\infty\}$ onto itself. See Jones and Singerman [1987, p. 17].) In this respect, $\mathbb{C} \cup \{\infty\}$ is like the euclidean plane, the cylinder and the torus, which also admit nontrivial complex functions (Section 2.10). Such surfaces are called *Riemann surfaces* and their general definition was first given by Weyl [1913].

There is an analogue of the Killing–Hopf theorem for Riemann surfaces called the *uniformization theorem*. The term "uniformization" comes from the original formulation of the theorem, proved by Poincaré [1907] and Koebe [1907], which states that any algebraic curve $p(x, y) = 0$ can be parameterized, or "uniformized", by complex functions $x = f(z)$, $y = g(z)$. Weyl's formulation is that any Riemann surface is the quotient of a simply connected Riemann surface by a discontinuous, fixed point free group, and that the simply connected Riemann surfaces are just $\mathbb{C}$, $\mathbb{C} \cup \{\infty\}$, and the hyperbolic plane (the subject of Chapter 4).

4

The Hyperbolic Plane

4.1 Negative Curvature and the Half-Plane

Historically speaking, the hyperbolic plane was the result of the search for a *non-euclidean plane*—a surface with unbounded straight lines and, for each line L and point $P \notin L$, more than one line through P which does not meet L. Such a surface departs from the euclidean plane in the opposite way to the sphere, and the hyperbolic plane, in fact, emerged from the study of surfaces which "curve" in the opposite way to the sphere. The train of thought, in brief, was this.

The sphere has the property that the curves obtained by cutting it by normal planes at a point P all have their radii of curvature on the same side (namely, ending at the center of the sphere). On the other hand, a saddle-shaped surface (Figure 4.1) has sections K_1, K_2 with their radii ρ_1, ρ_2 of curvature on opposite sides. (The radius of curvature of a plane curve K at P is the radius of the circle which most closely approximates K at P.) The curvature of a section of radius ρ is measured by $1/\rho$; hence, a good measure of the "combined curvature" of sections of radii ρ_1, ρ_2 is $1/\rho_1\rho_2$, where the radii are signed to distinguish their possible directions. When the sections are chosen so that ρ_1, ρ_2 take the maximum and minimum signed values at P, then $\kappa = 1/\rho_1\rho_2$ is called the *Gaussian curvature* at P. Thus, the Gaussian curvature of a sphere of radius ρ is the positive constant $1/\rho^2$, whereas the Gaussian curvature of a saddle-shaped piece of surface is negative at all points. From now on we simply call κ the *curvature*.

The curvature of the euclidean plane is, of course, 0, as is the curvature of a cylinder because one of its sections of extreme curvature is a straight line, which has curvature 0. We should expect the hyperbolic plane, if it exists, to be a surface of constant negative curvature. However, the first surface of constant negative curvature actually discovered was not a plane, but the analogue of a cylinder. For historical reasons it is called the *pseudosphere*.

The pseudosphere is the surface of revolution of the *tractrix*, a curve which is most conveniently defined as the *involute* of the catenary $v = \cosh u$ in the (u, v)-plane. Intuitively speaking, the involute is the path of the end of a piece of string being unwound from the catenary. That is, it is the locus of the endpoint Q of the tangent PQ whose length equals the length of the catenary from S to $P = (\sigma, \cosh \sigma)$ (Figure 4.2).

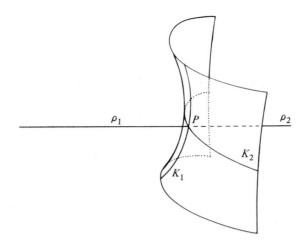

FIGURE 4.1.

Conversely, the catenary's right half is the locus of the center of curvature P of the tractrix. This fact is the key to the constant curvature of the pseudosphere, as we shall see in the proof of the theorem below. Also, when one searches for coordinates on the pseudosphere which make its distance function as simple as possible, one finds a distance function on the upper half-plane—a surface which *is* a plane, topologically—which makes it locally isometric to the pseudosphere. In this way, the hyperbolic plane emerges unexpectedly as the upper half-plane $\mathbb{H}^2 = \{(x,y) \mid y > 0\}$, with a non-euclidean distance function.

Theorem. *The pseudosphere has constant negative curvature* -1 *and is locally isometric to* $\mathbb{H}^2$ *with distance function defined by* $ds^2 = \frac{dx^2+dy^2}{y^2}$.

Proof. Using the fact that $\frac{dv}{du} = \sinh u$, one easily calculates that, in Figure 4.2,

$$PQ = \text{arc } PS = \int_0^\sigma \sqrt{1 + \left(\frac{dv}{du}\right)^2}\, du = \sinh \sigma$$

and that $R = (\sigma - \coth \sigma, 0)$. It follows that

$$PR = \frac{\cosh^2 \sigma}{\sinh \sigma}$$

and hence

$$QR = \frac{\cosh^2 \sigma - \sinh^2 \sigma}{\sinh \sigma} = \frac{1}{\sinh \sigma} = \frac{1}{PQ}. \tag{1}$$

But it is clear from Figure 4.2 that PQ and QR are the radii of curvature at Q of two normal sections of the pseudosphere, the first by the (u, v)-plane and the second by the plane normal to it through PQ. It is also

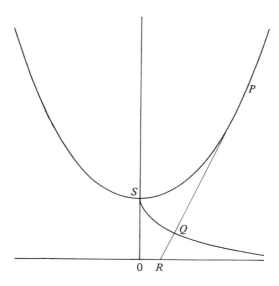

FIGURE 4.2.

clear, by symmetry, that these sections give the maximum and minimum radii of curvature among all normal sections at P; and that these radii have opposite sign. Hence,

$$\text{Gaussian curvature} = -\frac{1}{PQ} \cdot \frac{1}{QR} = -1 \quad \text{by (1)}.$$

Now to find simpler coordinates on the pseudosphere, we first compute the coordinates of Q in Figure 4.2, obtaining the following parametric equations for the tractrix:

$$u = \sigma - \tanh \sigma, \quad v = \operatorname{sech} \sigma. \tag{2}$$

A more natural parameter to use in place of σ is the arc length τ along the tractrix:

$$\tau = \int_0^\sigma \sqrt{du^2 + dv^2} = \log \cosh \sigma. \tag{3}$$

by (2). Thus, $\cosh \sigma = e^\tau$ and hence by (2) again

$$v = e^{-\tau}. \tag{4}$$

We now take τ and the angle x (Figure 4.3) as coordinates on the pseudo-sphere. Then the length subtended by angle dx on a circular cross section is $v\,dx = e^{-\tau}dx$ by (4), and hence the infinitesimal distance ds between points with coordinates (x, τ) and $(x + dx, \tau + d\tau)$ is given by

$$ds^2 = e^{-2\tau}\,dx^2 + d\tau^2. \tag{5}$$

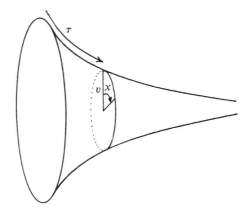

FIGURE 4.3.

Finally we introduce the variable $y = e^\tau$ so that $dy = e^\tau d\tau$ and (5) simplifies to

$$ds^2 = e^{-2\tau}(dx^2 + dy^2) = \frac{dx^2 + dy^2}{y^2}. \tag{6}$$

Thus, the pseudosphere is locally isometric to the (x, y)-plane if distance in the latter is defined by (6). Only the upper half-plane $\mathbb{H}^2$ is relevant since $y = e^\tau > 0$ for all $\tau \in \mathbb{R}$.   □

Remarks. (1) It is nice to have the pseudosphere in $\mathbb{R}^3$ as a concrete surface of constant negative curvature because it enables us to interpret hyperbolic geometry as the geometry of euclidean distance on a curved surface. However, the pseudosphere is less satisfactory than its positive curvature counterpart $\mathbb{S}^2$ in several respects. First, it is not the "plane" we are looking for, but rather a cylinder. Second, it is not complete because the tractrix ends at $u = 0$. Third, only some of the "lines" on the pseudosphere (namely, the tractrix sections) have a simple description in $\mathbb{R}^3$. In fact, Hilbert [1901] showed that any surface of constant negative curvature smoothly embedded in $\mathbb{R}^3$ is incomplete. This leads us to abandon the euclidean view of negative curvature and take up the alternative offered by the theorem—a complete, topologically planar surface $\mathbb{H}^2$ with a non-euclidean distance function.

(2) We shall see in the next section that the loss of the euclidean notion of distance is not a great blow to geometric intuition. In fact, it is possible to find viewpoints from which important geometric aspects of $\mathbb{H}^2$ "look euclidean". One that *always* looks euclidean is angle. Since the infinitesimal distance

$$ds = \frac{\sqrt{dx^2 + dy^2}}{y}$$

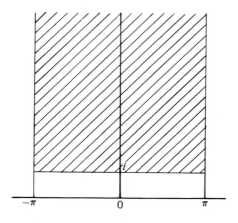

FIGURE 4.4.

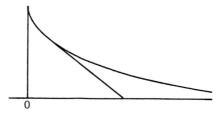

FIGURE 4.5.

in $\mathbb{H}^2$ is simply the euclidean distance $\sqrt{dx^2 + dy^2}$ divided by y, the ratio between the two is a constant, independent of direction. This means that angle, which is determined by the ratio of side lengths of infinitesimal triangles, is the same when determined by either distance function.

Exercises

4.1.1. Check the calculations behind (1), (2), and (3).

4.1.2. Show that the pseudosphere corresponds to the portion of $\mathbb{H}^2$ shown shaded in Figure 4.4.

4.1.3. Show that the distance between tractrix sections of the pseudosphere tends to 0 exponentially with respect to distance along the tractrix.

4.1.4. Prove the geometric property of the tractrix illustrated in Figure 4.5: the length of the tangent above the u-axis is constant.

4.1.5. Define a homeomorphism from the plane to the upper half plane.

4.2 The Half-Plane Model and the Conformal Disc Model

Studying the hyperbolic plane in terms of $\mathbb{H}^2$ is rather like studying $\mathbb{S}^2$ in terms of its stereographic image $\mathbb{C} \cup \{\infty\}$, except that there is no "real" hyperbolic plane to fall back on. $\mathbb{H}^2$ with $ds = \sqrt{dx^2 + dy^2}/y$ *is* the hyperbolic plane, as much as anything can be, but we call it a *model* of the hyperbolic plane because any surface isometric to $\mathbb{H}^2$ is equally entitled to the name. In fact, there are several models of the hyperbolic plane in common use and one in particular, called the $\mathbb{D}^2$-model, complements $\mathbb{H}^2$ particularly well. We shall introduce the $\mathbb{D}^2$-model later in this section, after we have exposed some inconvenient features of $\mathbb{H}^2$.

We define the $\mathbb{H}^2$-*model* to be $\mathbb{H}^2$ with the local distance function given by $ds = \sqrt{dx^2 + dy^2}/y$. $\mathbb{H}^2$ is certainly homeomorphic to the euclidean plane $\mathbb{R}^2$ though presumably not isometric to it. So far, all we know of the geometry of $\mathbb{H}^2$ is that its angles are the same as euclidean angles [Remark (2) of Section 4.1]. To find other geometric properties of $\mathbb{H}^2$ we have to follow Klein's program of deriving geometry from the properties of isometries.

As in euclidean and spherical geometry, isometries of $\mathbb{H}^2$—or $\mathbb{H}^2$-*isometries* as we shall call them—are most simply expressed as complex functions, in this case functions of $z = x + iy$. In terms of z, the infinitesimal distance $ds = \sqrt{dx^2 + dy^2}/y$ is $|dz|/\text{Im } z$, and some functions $\mathbb{H}^2 \to \mathbb{H}^2$ which obviously leave ds invariant are

   (i) $t_\alpha(z) = \alpha + z$ for any $\alpha \in \mathbb{R}$,

   (ii) $d_\rho(z) = \rho z$ for any positive $\rho \in \mathbb{R}$, and

   (iii) $\bar{r}_{OY}(z) = -\bar{z}$ (reflection in the y-axis).

These functions are familiar to us already as euclidean translations, dilatations, and reflection respectively. In fact, every $\mathbb{H}^2$-isometry, when conjugated to a suitable position, becomes a euclidean mapping. This is very helpful in visualizing hyperbolic geometry, though one has to remember euclidean dilatations and translations have a different geometric meaning in $\mathbb{H}^2$. A euclidean dilatation d_ρ ($\rho \neq 1$) is not an $\mathbb{H}^2$-dilatation (there are none). And a euclidean translation t_α ($\alpha \neq 0$) is not an $\mathbb{H}^2$-translation since, as we shall see, it leaves no "$\mathbb{H}^2$-lines" invariant. In fact, d_ρ is an $\mathbb{H}^2$-translation (with the y-axis as the invariant $\mathbb{H}^2$-line) and t_α is a type of isometry peculiar to the hyperbolic plane.

The "obvious" $\mathbb{H}^2$-isometries (i), (ii), (iii) are not enough to generate all $\mathbb{H}^2$-isometries. In particular, we seem to be lacking rotations. $\mathbb{H}^2$-rotations are best grasped by mapping $\mathbb{H}^2$ onto the open unit disc $\mathbb{D}^2$, where some rotations materialize as euclidean rotations about the origin.

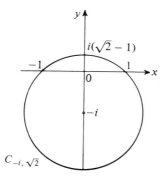

FIGURE 4.6.

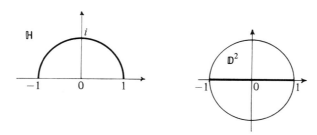

FIGURE 4.7.

To map $\mathbb{H}^2$ onto $\mathbb{D}^2$ we invert the z-plane in the circle $C_{-i,\sqrt{2}}$ (Figure 4.6) and then reflect in the x-axis. The inversion exchanges O and i because the product of their distances from the center $-i$ is $2 \times 1 = (\text{radius})^2$; so after the subsequent reflection in the x-axis. O ends up at $-i$. This gives an image of $\mathbb{H}^2$ in $\mathbb{D}^2$ the "right way up" (± 1 preserved, O sent the bottom, $-i$, and ∞ to the top, i). Also, by the geometric properties of inversion (Section 3.4), circles and signed angles are preserved. In particular, the euclidean semicircle in $\mathbb{H}^2$ through -1, i, 1 goes to the segment of x-axis in $\mathbb{D}^2$ (Figure 4.7). Since inversion in $C_{d,\rho}$ is given by the function $\frac{d\bar{z}+\rho^2-d\bar{d}}{\bar{z}-\bar{d}}$ (Section 3.5), our inversion is $\frac{-i\bar{z}+1}{\bar{z}-i}$ which, when followed by the reflection in the x-axis, gives

$$J(z) = \frac{iz+1}{z+i}$$

as the map of $\mathbb{H}^2$ onto $\mathbb{D}^2$.

We define the $\mathbb{D}^2$-*distance* between points $w_1, w_2 \in \mathbb{D}^2$ to be the $\mathbb{H}^2$-distance between their preimages $J^{-1}(w_1)$, $J^{-1}(w_2) \in \mathbb{H}^2$. Thus, the $\mathbb{D}^2$-*isometries* are the conjugates JhJ^{-1} of $\mathbb{H}^2$-isometries h. To verify that euclidean rotations about O are indeed $\mathbb{D}^2$-isometries, we work out a formula for the $\mathbb{H}^2$-distance $ds = |dz|/\operatorname{Im} z$ in terms of w.

Since $w = J(z) = \frac{iz+1}{z+i}$ we have $z = \frac{-iw+1}{w-i}$, hence

$$
\begin{aligned}
\frac{|dz|}{\operatorname{Im} z} &= \left| d\frac{-iw+1}{w-i} \right| \Big/ \operatorname{Im}\left(\frac{-iw+1}{w-i} \right) \\
&= \left| \frac{-2dw}{(w-i)^2} \right| \Big/ \operatorname{Im}\frac{(1-iw)(\bar{w}-\bar{i})}{(w-i)(\bar{w}-\bar{i})} \\
&= \frac{|2dw|}{|w-i|^2} \Big/ \operatorname{Im}\frac{(1-iw)(\bar{w}+i)}{|w-i|^2} \\
&= \frac{|2dw|}{(1-|w|^2)}.
\end{aligned}
$$

This is indeed invariant under euclidean rotations about O because they leave $|w|$ unchanged; hence we have the $\mathbb{D}^2$-isometries

(iv) $r_\theta(w) = e^{i\theta}w$ for $\theta \in \mathbb{R}$.

More generally, $|2dw|/(1-|w|^2)$ is invariant under any euclidean reflection in a line through O; in particular,

(v) $\bar{r}(w) = \bar{w}$.

It is appropriate to call (iv) and (v) $\mathbb{D}^2$-*rotations* (about O) and a $\mathbb{D}^2$-*reflection* (in the real axis), respectively, and their conjugates by J^{-1} will be $\mathbb{H}^2$-*rotations* (about i) and an $\mathbb{H}^2$-*reflection* (in the unit circle).

The $\mathbb{H}^2$-reflection $J^{-1}\bar{r}J$ is, in fact, inversion I in the unit circle because

$$
z = \frac{-iw+1}{w-i} \quad \overset{\bar{r}}{\mapsto} \quad \frac{-i\bar{w}+1}{\bar{w}-i} = \frac{-i\left(\frac{iz+1}{z+i}\right)+1}{\left(\frac{iz+1}{z+i}\right)-i} = \frac{1}{\bar{z}};
$$

thus, we have found that I is an $\mathbb{H}^2$-isometry. It follows by the $\mathbb{H}^2$-isometries (i) and (ii) that inversion $t_\alpha d_\rho I d_\rho^{-1} t_\alpha^{-1}$ in any circle $C_{\alpha,\rho}$ is also an $\mathbb{H}^2$-isometry. These inversions, together with the euclidean reflections $t_\alpha \bar{r}_{OY} t_\alpha^{-1}$ in euclidean lines $x = \alpha$, make up all the $\mathbb{H}^2$-reflections. In the next section we shall show that they generate all $\mathbb{H}^2$-isometries. Of more immediate interest is that they reveal the $\mathbb{H}^2$-*lines*—the fixed point sets of $\mathbb{H}^2$-reflections—to be euclidean semicircles with centers on the x-axis, and euclidean half lines $x = \alpha$ (with $y > 0$).

It is easy to see (Exercises) that there is a unique $\mathbb{H}^2$-line between any two $z_1, z_2 \in \mathbb{H}^2$, and that $\mathbb{H}^2$-lines extend indefinitely (because $\mathbb{H}^2$-distance tends to infinity as a point approaches the real axis). Thus, $\mathbb{H}^2$-lines behave like "lines" in a "plane"; but not like euclidean lines, because for any $\mathbb{H}^2$-line L and point $P \notin L$ there is more than one $\mathbb{H}^2$-line through P not meeting L (see e.g., Figure 4.8). $\mathbb{H}^2$ is the long sought "non-euclidean plane".

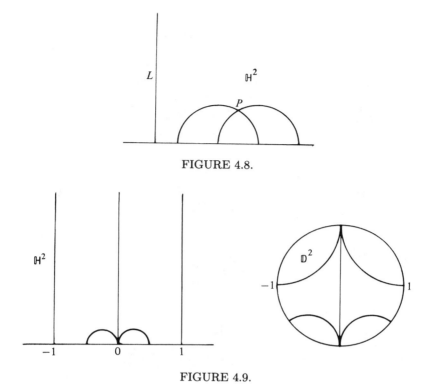

FIGURE 4.8.

FIGURE 4.9.

Passing back to $\mathbb{D}^2$, we find the $\mathbb{D}^2$-*lines* to be the circular arcs orthogonal to the unit circle bounding $\mathbb{D}^2$, again using the fact that inversion preserves circles and angles. The $\mathbb{D}^2$-lines, of course, include the diameters of $\mathbb{D}^2$, which are euclidean line segments. Some $\mathbb{H}^2$-lines and their images in $\mathbb{D}^2$ are compared in Figure 4.9. The $\mathbb{D}^2$-*reflections*, as one would expect, are inversions in $\mathbb{D}^2$-lines (Exercise).

To sum up, the $\mathbb{H}^2$-*model* of the hyperbolic plane is the upper half-z-plane $\mathbb{H}^2$ with distance $ds = |dz|/\text{Im } z$. The $\mathbb{H}^2$-*lines* are semicircles orthogonal to the real axis (including the upper half lines $\text{Re } z = $ constant) and $\mathbb{H}^2$-*angle* is the same as euclidean angle.

The $\mathbb{D}^2$-*model* is the open unit w-disc $\mathbb{D}^2$ with distance $ds = |2dw|/(1 - |w|^2)$. The $\mathbb{D}^2$-*lines* are circular arcs orthogonal to the boundary circle of $\mathbb{D}^2$ (including the diameters of $\mathbb{D}^2$) and $\mathbb{D}^2$-*angle* is the same as euclidean angle. (The last property is because the map $J : \mathbb{H}^2 \to \mathbb{D}^2$ is the product of two inversions, hence angle-preserving. Incidentally, this is why we call the $\mathbb{D}^2$-model "conformal"—to distinguish it from another disc model, considered in Section 4.8, which does *not* preserve angles.)

The splendid picture "Circle limit I" by M.C. Escher (from the Collection Haags Gemeentemuseum-The Hague) (Figure 4.10) gives a good impression

FIGURE 4.10. Picture by M.C. Escher. Used with permission of the Collection Haags Gemeentemuseum—The Hague.

of the $\mathbb{D}^2$-model. The fish are all of equal $\mathbb{D}^2$-size and are placed along $\mathbb{D}^2$-lines.

Exercises

4.2.1. Show that the $\mathbb{H}^2$-length of the segment of the y-axis from i to $i\beta$ is $|\log \beta|$.

4.2.2. Check directly that inversion in the unit circle is an $\mathbb{H}^2$-isometry.

4.2.3. Give a euclidean geometric construction showing existence and uniqueness of an $\mathbb{H}^2$-line through any $z_1, z_2 \in \mathbb{H}^2$.

4.2.4. Defining a circle, as usual, as the set of points equidistant from a point, find the $\mathbb{D}^2$-circles with $\mathbb{D}^2$-center O. Hence, show that $\mathbb{H}^2$-circles with $\mathbb{H}^2$-center i, and all other $\mathbb{H}^2$-circles, are all the euclidean circles in $\mathbb{H}^2$ (though their $\mathbb{H}^2$-centers are not their euclidean centers).

4.2.5. Show that the $\mathbb{D}^2$-radius of a euclidean circle of radius $r < 1$ and center O is $2\int_0^r \frac{dx}{1-x^2} = 2\tanh^{-1} r$. Deduce that the $\mathbb{D}^2$-circumference of a $\mathbb{D}^2$-circle of $\mathbb{D}^2$-radius ρ is $2\pi\sinh\rho$.

4.2.6. Show (by elementary geometry) that the $\mathbb{S}^2$-circumference of an $\mathbb{S}^2$-circle of $\mathbb{S}^2$-radius ρ is $2\pi\sin\rho$.

4.3   The Three Reflections Theorem

We do not yet know whether we have found all $\mathbb{H}^2$-isometries (or $\mathbb{D}^2$-isometries); however, we have found enough to be able to bring any point or line segment to convenient position. By choice of positions which make calculations easier, we shall now derive the basic properties of $\mathbb{H}^2$-distance. Our aim is to prove the $\mathbb{H}^2$ form of the lemma that was fundamental to the classification of euclidean and spherical isometries—the one saying that the equidistant set of two points is a line, and that reflection in the equidistant line exchanges the two points.

Our first proposition depends on moving an arbitrary $\mathbb{H}^2$-line segment PQ in $\mathbb{H}^2$ to the y-axis by $\mathbb{H}^2$-isometries. This can be done by moving P onto the y-axis by a t_α, then rotating about P until Q is on the y-axis (i.e., until the image of Q in $\mathbb{D}^2$ is on the y-axis). Since these $\mathbb{H}^2$-isometries preserve circles and angles, the $\mathbb{H}^2$-line segment between P and Q must now be part of a circle orthogonal to the real axis and passing through P and Q. This "circle" is the y-axis itself.

Proposition. *The $\mathbb{H}^2$-line segment between P and Q is the curve of shortest $\mathbb{H}^2$-length between P and Q.*

Proof. By the remarks above, we can assume that the $\mathbb{H}^2$-line segment PQ is a segment of the y-axis. Now if C is any other curve from P to Q (Figure 4.11) we have $\mathbb{H}^2$-length of $C = \int_C \frac{\sqrt{dx^2+dy^2}}{y} \geq \int_P^Q \frac{dy}{y} = \mathbb{H}^2$-length of PQ. $\qquad\square$

Corollary (triangle inequality). *If $P, Q, R \in \mathbb{H}^2$ then*

$$(\mathbb{H}^2\text{-length of } PR) + (\mathbb{H}^2\text{-length of } RQ) \geq (\mathbb{H}^2\text{-length of } PQ)$$

with strict inequality when R is not on the $\mathbb{H}^2$-line through P, Q.

Proof. The $\geq$ sign follows by the proposition, taking $PR \cup RQ$ as the curve C. If R is not on the $\mathbb{H}^2$-line through P, Q, which we again take to be on the y-axis, then we can assume a situation like that shown in Figure 4.12. (If, in fact, R is lower than P, the proof is similar, but easier.) Each

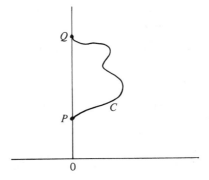

FIGURE 4.11.

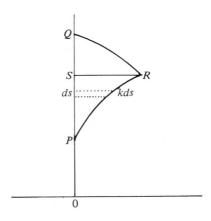

FIGURE 4.12.

infinitesimal segment of PR is k times the $\mathbb{H}^2$-length of the corresponding segment of PS, where $k \geq \sec(\text{angle } RPQ)$. Thus, k is strictly greater than 1, and hence

$$\mathbb{H}^2\text{-length of } PR > \mathbb{H}^2\text{-length of } PS.$$

Similarly,

$$\mathbb{H}^2\text{-length of } RQ > \mathbb{H}^2\text{-length of } SQ,$$

and, therefore,

$$(\mathbb{H}^2\text{-length of } PR) + (\mathbb{H}^2\text{-length of } RQ) > (\mathbb{H}^2\text{-length of } PQ). \qquad \square$$

Now to prove the important lemma we want to move arbitrary $P, P' \in \mathbb{H}^2$ to positions which are mirror images of each other in the y-axis. This can be done by first rotating about P until P' has the same y-coordinate as P. (Such a position must exist by the intermediate value theorem. We know

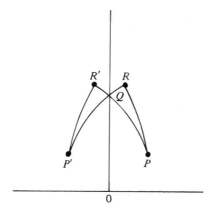

FIGURE 4.13.

that continuous rotation about O in $\mathbb{D}^2$ is possible, and this translates to continuous rotation in $\mathbb{H}^2$, taking P' from above P to below P.) Then one applies a suitable t_α to make P, P' (euclidean) equidistant from the y-axis.

Lemma. *The set of points $\mathbb{H}^2$-equidistant from two points $P, P' \in \mathbb{H}^2$ is an $\mathbb{H}^2$-line L, and $\mathbb{H}^2$-reflection in L exchanges P and P'.*

Proof. By the remarks above we can choose P, P' to be mirror images in the y-axis, so that reflection $\bar{r}_{OY}$ in the y-axis exchanges P and P'. Since $\bar{r}_{OY}$ is an $\mathbb{H}^2$-isometry which fixes each point Q on the y-axis, it follows that any such Q is $\mathbb{H}^2$-equidistant from P, P'. Thus, the $\mathbb{H}^2$-equidistant set of P, P' includes the y-axis—which is an $\mathbb{H}^2$-line.

Now suppose that the $\mathbb{H}^2$-equidistant set of P, P' includes a point R *not* on the y-axis. Then the mirror image R' of R is also $\mathbb{H}^2$-equidistant from P, P' (as the reflection $\bar{r}_{OY}$ sending R to R' is an $\mathbb{H}^2$-isometry which exchanges P and P') and we can assume, without loss of generality, a situation like that in Figure 4.13. We have

$$\mathbb{H}^2\text{-length of } P'R' = \mathbb{H}^2\text{-length of } PR \quad \text{by reflection}$$
$$= \mathbb{H}^2\text{-length of } P'R \quad \text{by hypothesis}$$
$$= (\mathbb{H}^2\text{-length of } P'Q) + (\mathbb{H}^2\text{-length of } QR)$$
$$= (\mathbb{H}^2\text{-length of } P'Q) + (\mathbb{H}^2\text{-length of } QR') \quad \text{by reflection,}$$

contrary to the strict triangle inequality. This contradiction shows that the y-axis is the complete $\mathbb{H}^2$-equidistant set of P, P'. $\square$

We have now done all the work needed to show that all $\mathbb{H}^2$-isometries are products of $\mathbb{H}^2$-reflections.

Theorem. *Each $\mathbb{H}^2$-isometry is the product of one, two, or three $\mathbb{H}^2$-reflections.*

Proof. As in the euclidean and spherical cases, one first observes that each isometry is determined by its effect on three points A, B, C not in a line. This holds in $\mathbb{H}^2$ because each point $P \in \mathbb{H}^2$ is determined by its $\mathbb{H}^2$-distances from A, B, C not in an $\mathbb{H}^2$-line—if there were two points P, P' with the same $\mathbb{H}^2$-distances from A, B, C then A, B, C would lie in the $\mathbb{H}^2$-equidistant set of P, P', contrary to the lemma above.

Now to express a given $\mathbb{H}^2$-isometry f as the product of $\mathbb{H}^2$-reflections, one chooses any $A, B, C \in \mathbb{H}^2$ not in an $\mathbb{H}^2$-line and uses one, two, or three $\mathbb{H}^2$-reflections to send A, B, C to $f(A)$, $f(B)$, $f(C)$, respectively, using the exchange of equidistant points by reflections exactly as in the euclidean and spherical cases. $\qquad\square$

Corollary. *The $\mathbb{H}^2$-isometries form a group.*

Proof. As in the euclidean and spherical cases because each reflection is self-inverse. $\qquad\square$

Exercises

4.3.1. Show, by suitable choice of axes, that inverses P, P' with respect to any circle $C \subset \mathbb{C}$ can be viewed as $\mathbb{H}^2$-mirror images in an $\mathbb{H}^2$-line C. Deduce that inverses with respect to a circle are preserved by inversion (in any other circle).

4.3.2. Deduce from Exercise 4.3.1 that $\mathbb{D}^2$-reflections are inversions in $\mathbb{D}^2$-lines, and hence that each $\mathbb{D}^2$-isometry is the product of one, two, or three inversions.

4.3.3. Show that each t_α and d_ρ is the product of two $\mathbb{H}^2$-reflections.

4.3.4. Show that $\mathbb{H}^2$-isometries map $\mathbb{H}^2$-lines onto $\mathbb{H}^2$-lines (and similarly for $\mathbb{D}^2$).

4.4   Isometries as Complex Functions

Having established that all $\mathbb{H}^2$-isometries are products of inversions, in circles $C_{\alpha,\rho}$ or lines $x = \alpha$, we can use the formula for inversion (Section 3.5) to express $\mathbb{H}^2$-isometries as complex functions. As in the euclidean and spherical cases, there is a partition of isometries into *orientation-preserving* (products of an even number of reflections) and *orientation-reversing* (products of an odd number of reflections). The orientation-preserving isometries obviously form a subgroup $\mathrm{Iso}^+(\mathbb{H}^2)$ of the group $\mathrm{Iso}(\mathbb{H}^2)$ of all $\mathbb{H}^2$-isometries. The orientation-reversing elements form the coset $\mathrm{Iso}^+(\mathbb{H}^2) \cdot \bar{r}_{OY}$, where $\bar{r}_{OY}$ is the reflection in the y-axis. We shall see that $\bar{r}_{OY} \notin \mathrm{Iso}^+(\mathbb{H}^2)$ as soon as we find the form of elements of $\mathrm{Iso}^+(\mathbb{H}^2)$ because $\bar{r}_{OY}$ has a line of fixed points and it will transpire that each non-trivial element of $\mathrm{Iso}^+(\mathbb{H}^2)$ has at most two fixed points.

Theorem (Poincaré [1882]). *The $\mathbb{H}^2$-isometries are of the form*

$$f(z) = \frac{\alpha z + \beta}{\gamma z + \delta},$$

where $\alpha, \beta, \gamma, \delta \in \mathbb{R}$ and $\alpha\delta - \beta\gamma = 1$ (orientation-preserving) and

$$\bar{f}(z) = \frac{-\alpha \bar{z} + \beta}{-\gamma \bar{r} + \delta},$$

where $\alpha, \beta, \gamma, \delta \in \mathbb{R}$ and $\alpha\delta - \beta\gamma = 1$ (orientation-reversing).

Proof. Inversion in $C_{\epsilon,\rho}$ is $\bar{g}(z) = \frac{\epsilon\bar{r} + \rho^2 - \epsilon^2}{\bar{r} - \epsilon}$ (by Section 3.5), which has determinant $-\epsilon^2 - \rho^2 + \epsilon^2 = -\rho^2$. Hence, by multiplying numerator and denominator by $1/\rho$, we get determinant -1. Likewise, reflection $t_\epsilon \bar{r}_{OY} t_\epsilon^{-1}$ in the line $x = \epsilon$ is the function

$$\bar{h}(z) = \epsilon - (-\overline{\epsilon + z}) = -\bar{z} + 2\epsilon = \frac{-\bar{z} + 2\epsilon}{1},$$

which has determinant -1. Hence, the product of any two $\mathbb{H}^2$-reflections is a function of the form

$$f(z) = \frac{\alpha z + \beta}{\gamma z + \delta} \quad \text{with determinant } \alpha\delta - \beta\gamma = 1.$$

And the product of such a function with another $\mathbb{H}^2$-reflection will be of the form

$$\bar{f}(z) = \frac{\alpha' \bar{z} + \beta'}{\gamma' \bar{z} + \delta'} \quad \text{with determinant } \alpha'\delta' - \beta'\gamma' = -1,$$

which is the same thing as

$$\bar{f}(z) = \frac{-\alpha\bar{z} + \beta}{-\gamma\bar{z} + \delta} \quad \text{with } \alpha\delta - \beta\gamma = 1.$$

By the three reflections theorem (Section 4.3), this covers all $\mathbb{H}^2$-isometries.

Conversely, given a function $f(z) = \frac{\alpha z + \beta}{\gamma z + \delta}$ with $\alpha\delta - \beta\gamma = 1$, we write it in the form

$$f(z) = \frac{\alpha z + (\alpha\delta/\gamma) + \beta - (\alpha\delta/\gamma)}{\gamma z + \delta}$$

$$= \frac{\alpha}{\gamma} + \frac{\beta - \alpha\delta/\gamma}{\gamma z + \delta}$$

$$= \frac{\alpha}{\gamma} + \frac{\beta\gamma - \alpha\delta}{\gamma(\gamma z + \delta)} = \frac{\alpha}{\gamma} - \frac{1}{\gamma(\gamma z + \delta)}.$$

Assuming for the moment that $\gamma > 0$, this function is a product of the $\mathbb{H}^2$-isometries $z \mapsto \gamma z$, $z \mapsto z+\epsilon$ for $\epsilon = \delta$, α/γ and $z \mapsto -1/z$ (the product of $z \mapsto 1/\bar{z}$ and $z \mapsto -\bar{z}$), hence itself an $\mathbb{H}^2$-isometry. If $\gamma < 0$, we write $f(z) = \frac{-\alpha z - \beta}{-\gamma z - \delta}$, which still has determinant 1, and repeat the argument. Finally, if $\gamma = 0$, we simply have $f(z) = \frac{\alpha}{\delta}z + \frac{\beta}{\delta}$ with $\alpha\delta = 1$, so we also have $\alpha/\delta > 0$; hence, this is the $\mathbb{H}^2$-isometry $t_{\beta/\delta}d_{\alpha/\delta}$.

Notice that in each case f is orientation-preserving because each t_ϵ or d_ρ is the product of two $\mathbb{H}^2$-reflections (by Exercise 4.3.3).

Finally, given $\bar{f}(z) = \frac{-\alpha\bar{z}+\beta}{-\gamma\bar{z}+\delta}$, we write it as the product of $f(z) = \frac{\alpha z+\beta}{\gamma z+\delta}$ with $\bar{r}_{OY}(z) = -\bar{z}$. We know that $f(z)$ is an orientation-preserving $\mathbb{H}^2$-isometry; hence, $\bar{f}(z) = f\bar{r}_{OY}(z)$ is an orientation-reversing $\mathbb{H}^2$-isometry. $\qquad\square$

Corollary 1. *The orientation-preserving and orientation-reversing isometries form complementary sets.*

Proof. By the remarks preceding the theorem, it will suffice to show that $\bar{r}_{OY}$ is not orientation-preserving. This is so because $\bar{r}_{OY}$ has the y-axis as its fixed point set, whereas each orientation-preserving $\mathbb{H}^2$-isometry has at most two fixed points, namely, the solutions of the quadratic equation

$$z = \frac{\alpha z + \beta}{\gamma z + \delta}.$$ $\qquad\square$

Corollary 2. *The $\mathbb{D}^2$-isometries are the functions*

$$f(z) = \frac{az + b}{\bar{b}z + \bar{a}}$$

where $a, b \in \mathbb{C}$ and $|a|^2 - |b|^2 = 1$ (orientation-preserving) and

$$\bar{f}(z) = \frac{a\bar{z} + b}{\bar{b}\bar{z} + \bar{a}}$$

where $a, b \in \mathbb{C}$ and $|a|^2 - |b|^2 = 1$ (orientation-reversing).

Proof. We know from Section 4.2 that the $\mathbb{D}^2$-isometries are the functions JhJ^{-1} where h is an $\mathbb{H}^2$-isometry and

$$J(z) = \frac{iz + 1}{z + i}, \quad J^{-1}(z) = \frac{-iz + 1}{z - i}.$$

If h is an orientation-preserving $\mathbb{H}^2$-isometry $h(z) = \frac{\alpha z+\beta}{\gamma z+\delta}$, then JhJ^{-1} has matrix

$$\begin{pmatrix} i & 1 \\ 1 & i \end{pmatrix}\begin{pmatrix} \alpha & \beta \\ \gamma & \delta \end{pmatrix}\begin{pmatrix} -i & 1 \\ 1 & -i \end{pmatrix} = \begin{pmatrix} \alpha + i\beta - i\gamma + \delta & i\alpha + \beta + \gamma - i\delta \\ -i\alpha + \beta + \gamma + i\delta & \alpha - i\beta + i\gamma + \delta \end{pmatrix}$$

with determinant

$$\det \begin{pmatrix} i & 1 \\ 1 & i \end{pmatrix} \det \begin{pmatrix} \alpha & \beta \\ \gamma & \delta \end{pmatrix} \det \begin{pmatrix} -i & 1 \\ 1 & -i \end{pmatrix} = (-2)1(-2) = 4.$$

The matrix is of the form

$$\begin{pmatrix} c & d \\ \bar{d} & \bar{c} \end{pmatrix}$$

where $c = \alpha + i\beta - i\gamma + \delta$, $d = i\alpha + \beta + \gamma - i\delta$ and, hence, $4 = \det = c\bar{c} - d\bar{d} = |c|^2 - |d|^2$. We can, therefore, write

$$f(z) = JhJ^{-1}(z) = \frac{az+b}{\bar{b}z+\bar{a}} \quad \text{with } |a|^2 - |b|^2 = 1$$

by taking $a = c/2$, $b = d/2$.

Conversely, any function of this form is obtainable by suitable choice of α, β, γ, δ. Given any $a, b \in \mathbb{C}$ with $|a|^2 - |b|^2 = 1$, we form $c = 2a$, $d = 2b$ and solve the equations $c = \alpha + i\beta - i\gamma + \delta$, $d = i\alpha + \beta + \gamma - i\delta$ for α, β, γ, δ. The solutions are

$$\alpha = \operatorname{Re} a + \operatorname{Im} b, \quad \beta = \operatorname{Re} b + \operatorname{Im} a, \quad \gamma = \operatorname{Re} b - \operatorname{Im} a, \quad \delta = \operatorname{Re} a - \operatorname{Im} b,$$

and one readily checks that $\alpha\delta - \beta\gamma = |a|^2 - |b|^2 = 1$.

Finally, to produce the orientation-reversing $\mathbb{D}^2$-isometry $\bar{f}(z) = \frac{a\bar{z}+b}{\bar{b}z+\bar{a}}$ with $|a|^2 - |b|^2 = 1$, we simply compose the $f(z)$ obtained above with the orientation-reversing $\mathbb{D}^2$-isometry $\bar{r}(z) = \bar{z}$. $\qquad \square$

Exercises

4.4.1. Observing that $w \mapsto \frac{w-w_1}{-\bar{w}_1 w+1}$ is a $\mathbb{D}^2$-isometry that sends w_1 to O, deduce from Exercise 4.2.5 that the $\mathbb{D}^2$-distance between $w_1, w_2 \in \mathbb{D}^2$ is

$$2\tanh^{-1} \left| \frac{w_2 - w_1}{1 - \bar{w}_1 w_2} \right|.$$

4.4.2. Show (by elementary geometry) that the points 0, $\alpha \in \mathbb{R}$ in $\mathbb{C}$ are stereographic images of points on $\mathbb{S}^2$ whose $\mathbb{S}^2$-distance apart is $2\tan^{-1}\alpha$.

4.4.3. Using the mapping $w \mapsto \frac{w-w_1}{\bar{w}_1 w+1}$, induced by a rotation of $\mathbb{S}^2$, to bring w_1 to O, deduce that the $\mathbb{S}^2$-distance between points whose stereographic images are w_1, w_2 is

$$2\tan^{-1} \left| \frac{w_2 - w_1}{1 + \bar{w}_1 w_2} \right|.$$

4.4.4. Show that $w \mapsto \frac{(1-i\sigma)w+i\sigma}{-i\sigma w+(1+i\sigma)}$ is a $\mathbb{D}^2$-isometry with 1 as its only fixed point (such a $\mathbb{D}^2$-isometry is called a *limit rotation about* 1 because 1 is a euclidean limit of points in $\mathbb{D}^2$, though not itself in $\mathbb{D}^2$).

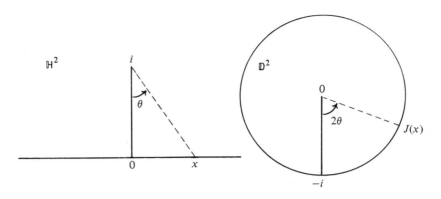

FIGURE 4.14.

4.4.5. Deduce from Exercise 4.4.4 that any $\mathbb{D}^2$-isometry is of the form

(rotation about 0)(limit rotation about 1)(rotation about 0).

4.4.6. Show that J maps the real axis onto the unit circle in the manner shown in Figure 4.14.

4.5 Geometric Description of Isometries

An important asset of the hyperbolic plane is its natural boundary, called the *circle at infinity*. In the $\mathbb{D}^2$-model the circle at infinity $\partial \mathbb{D}^2$ is the unit circle, the set of points that are euclidean limits of points of $\mathbb{D}^2$, without themselves being points of $\mathbb{D}^2$. These points are at infinity in terms of $\mathbb{D}^2$-distance (as the Escher picture at the end of Section 4.2 clearly shows) since the $\mathbb{D}^2$-distance from O of a point w tends to infinity as $|w|$ tends to 1. The circle at infinity $\partial \mathbb{H}^2$ of $\mathbb{H}^2$ is J^{-1} of the unit circle, namely, $\mathbb{R} \cup \{\infty\}$.

Many concepts of hyperbolic geometry are most simply described in terms of the circle at infinity. For example, we can define *asymptotic lines* to be lines with a common point on the circle at infinity (although it should be stressed that this point does not belong to the lines; rather, it is a common *limit point* or *end* of the lines). Asymptotic lines are thus distinguished from *ultraparallel* lines, which have no common point, either in the hyperbolic plane itself or on the circle at infinity (Figure 4.15). Corresponding to this distinction we have a distinction between the product of reflections in ultraparallel lines and the product of reflections in asymptotic lines. The former is a *translation*, whereas the latter is a *limit rotation*, because it is analogous to a rotation about the common limit point of the two lines of reflection. (For another reason for regarding this as a rotation, see Exercise 4.5.1.)

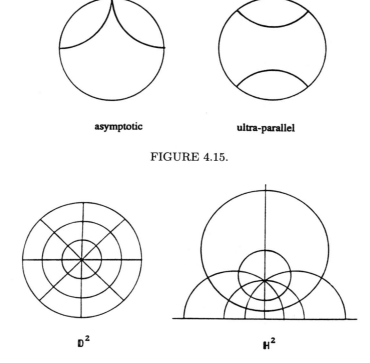

asymptotic   ultra-parallel

FIGURE 4.15.

$\mathbb{D}^2$ $\mathbb{H}^2$

FIGURE 4.16.

As we know, there are also *rotations* (products of reflections in pairs
of properly intersecting lines), and it is easy to see that there are *glide
reflections*. Before establishing that these four types are the only hyperbolic
isometries, we look at them in more detail. Taking advantage of coordinates
which make them "look euclidean", and moving between $\mathbb{H}^2$ and $\mathbb{D}^2$, we can
easily see how they affect the points and lines of the plane. In each case, the
isometry permutes the lines of a certain family and leaves invariant each
curve orthogonal to all the lines of the family.

(i) Rotation

Rotation about O in $\mathbb{D}^2$ is $r_\theta(w) = e^{i\theta}w$. This permutes the diameters of
$\mathbb{D}^2$ (the $\mathbb{D}^2$-lines through O) and leaves invariant the circles with center
O (the $\mathbb{D}^2$-circles with $\mathbb{D}^2$-center O). Figure 4.16 shows this and also the
image in $\mathbb{H}^2$. There is only one fixed point, the center of rotation (O in $\mathbb{D}^2$,
and i in $\mathbb{H}^2$).

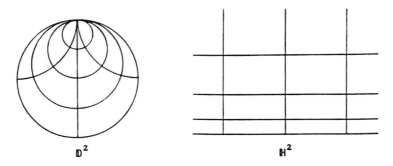

$\mathbb{D}^2$ $\mathbb{H}^2$

FIGURE 4.17.

(ii) Limit Rotation

Limit rotation about ∞ in $\mathbb{H}^2$ is $t_\alpha(z) = \alpha + z$. This permutes the lines $x = $ constant (the $\mathbb{H}^2$-lines through ∞) and leaves invariant the lines $y = $ constant. The latter curves, which are not $\mathbb{H}^2$-lines, are called *horocycles*. In $\mathbb{D}^2$, they are euclidean circles tangential to the circle at infinity (Figure 4.17). There is no fixed point in $\mathbb{D}^2$ or $\mathbb{H}^2$, but there is one on the circle at infinity (i for $\mathbb{D}^2$, ∞ for $\mathbb{H}^2$). It is the common end of the permuted lines.

(iii) Translation

Translation along the line from O to ∞ in $\mathbb{H}^2$ is $d_\rho(z) = \rho z$, where $\rho > 0$. This permutes the semicircles with center O ($\mathbb{H}^2$-lines perpendicular to the y-axis) and leaves invariant the y-axis (an $\mathbb{H}^2$-line called the *axis* of the translation) and the lines $y = $ constant $\times x$. The latter lines are not $\mathbb{H}^2$-lines, but *equidistant curves* of the y-axis. In $\mathbb{D}^2$ they are circles through $-i$ and i (Figure 4.18). There is no fixed point in $\mathbb{D}^2$ or $\mathbb{H}^2$, but there are two on the circle at infinity ($\pm i$ for $\mathbb{D}^2$, 0 and ∞ for $\mathbb{H}^2$), at the ends of the invariant $\mathbb{H}^2$- or $\mathbb{D}^2$-line.

To see why the euclidean line $y = $ constant $\times x$ is $\mathbb{H}^2$-equidistant from the y-axis, consider the perpendicular line segments PQ, $P'Q'$ between them (Figure 4.19). PQ is obviously mapped onto $P'Q'$ by a dilatation, which is on $\mathbb{H}^2$-isometry; hence, all such segments have the same $\mathbb{H}^2$-length.

(iv) Glide Reflection

A glide reflection is the product of a reflection with translation whose axis is the line of reflection; hence, we can use the above description of translation if the line of reflection is taken to be the y-axis. Like a translation, a glide reflection has one invariant line and two fixed points, on the circle at infinity (at the ends of the invariant line).

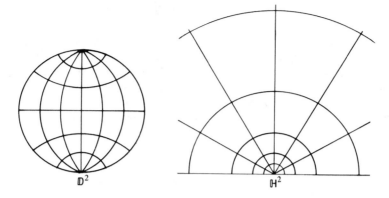

FIGURE 4.18.

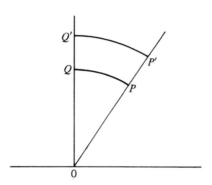

FIGURE 4.19.

Exercises

4.5.1. Show that the circular cross sections of the pseudosphere correspond to horocycles, and that a rotation of the pseudosphere about its axis of symmetry corresponds to a limit rotation.

4.5.2. Show that a limit rotation moves points through arbitrarily small distances.

4.5.3. Show that a translation moves points along equidistant curves further than it moves points along its axis and that the length of movement increases with the distance from the axis.

4.6   Classification of Isometries

As in euclidean and spherical geometry, glide reflection is the *only* type of orientation-reversing isometry. This can be proved as in Section 1.5 by converting any product of three reflections to $\bar{r}_N \bar{r}_M \bar{r}_L$, where L is the common perpendicular of lines M, N (see Exercises). However, there are more cases to consider because of the existence of asymptotic lines and ultraparallels. As a shortcut, we shall use the formula for the general orientation-reversing $\mathbb{H}^2$-isometry (Section 4.4) to find the invariant line L directly, or rather, its ends.

Lemma. *An orientation-reversing $\mathbb{H}^2$-isometry $\bar{f}$ has two fixed points on the circle at infinity.*

Proof. By Section 4.4 we can assume

$$\bar{f}(z) = \frac{-\alpha \bar{z} + \beta}{-\gamma \bar{z} + \delta}$$

where $\alpha, \beta, \gamma, \delta \in \mathbb{R}$ and $\alpha\delta - \beta\gamma = 1$. Hence $\bar{f}$ has fixed points on the circle at infinity $\mathbb{R} \cup \{\infty\}$ which are the real solutions of the equation

$$x = \frac{-\alpha x + \beta}{-\gamma x + \delta} \quad (\text{since } \bar{x} = x \text{ for } x \in \mathbb{R} \cup \{\infty\}).$$

If $\gamma = 0$, the solutions are $x = \beta/(\alpha+\delta), \infty$, and if $\gamma \neq 0$, we get a quadratic equation $\gamma x^2 - (\alpha + \delta)x + \beta = 0$ with solutions

$$
\begin{aligned}
x &= \frac{\alpha + \delta \pm \sqrt{(\alpha + \delta)^2 - 4\beta\gamma}}{2\gamma} \\
&= \frac{\alpha + \delta \pm \sqrt{\alpha^2 + \delta^2 - 2\alpha\delta + 4\alpha\delta - 4\beta\gamma}}{2\gamma} \\
&= \frac{\alpha + \delta \pm \sqrt{(\alpha - \delta)^2 + 4}}{2\gamma} \quad \text{since } \alpha\delta - \beta\gamma = 1.
\end{aligned}
$$

Since $(\alpha - \delta)^2 + 4 > 0$, these are two solutions. Hence, $\bar{f}$ has two fixed points on the circle at infinity. □

This lemma reduces the classification of orientation-reversing isometries to the classification of orientation-preserving isometries, which, in turn, reduces to the classification of pairs of lines, as we shall now see.

Classification of isometries. *Each isometry of the hyperbolic plane is either a*

(i) *rotation*,

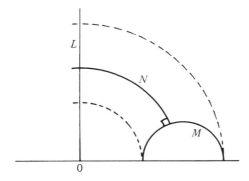

FIGURE 4.20.

(ii) *limit rotation,*

(iii) *translation, or*

(iv) *glide reflection.*

Proof. By Sections 4.3 and 4.4, each orientation-preserving isometry is the product of two reflections. The lines L, M of reflection can have only three essentially different positions:

(i) L, M intersect,

(ii) L, M are asymptotic, and

(iii) L, M are ultra-parallel.

Case (i). By suitable choice of coordinates, we can take the intersection of L, M to be O in $\mathbb{D}^2$. Then $\bar{r}_M \bar{r}_L$ is $r_\theta(w) = e^{i\theta} w$ for some $\theta \in \mathbb{R}$, i.e., a $\mathbb{D}^2$-rotation.

Case (ii). By suitable choice of coordinates, we can take L, M to be asymptotic at ∞ in $\mathbb{H}^2$, with L the y-axis. Then M is $x = \alpha/2$ for some $\alpha \in \mathbb{R}$ and $\bar{r}_M \bar{r}_L$ is $t_\alpha(z) = \alpha + z$, i.e., an $\mathbb{H}^2$-limit rotation.

Case (iii). By suitable choice of coordinates, we can take L to be the y-axis in $\mathbb{H}^2$ and M to be a semicircle disjoint from L. The various $\mathbb{H}^2$-lines perpendicular to L (Figure 4.20) make angles with M that vary continuously between 0 and π. Hence, by the intermediate value theorem, one of them, N, is a common perpendicular of L and M. Then, by a new choice of coordinates which makes N the y-axis, L and M become semicircles with center O (Figure 4.21). It is then clear that the product of $\mathbb{H}^2$-reflections in L, M is $d_\rho(z) = \rho z$ for some $\rho > 0$, i.e., an $\mathbb{H}^2$-translation.

Now if $\bar{f}$ is an orientation-reversing isometry the lemma gives us

Case (iv). $\bar{f}$ has two fixed points on the circle at infinity. Let L be the $\mathbb{H}^2$-line connecting the fixed points and consider the $\mathbb{H}^2$-isometry $f = \bar{f}\bar{r}_L$.

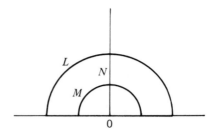

FIGURE 4.21.

Since $\bar{r}_L$ also fixes the ends of L, f is an orientation-preserving $\mathbb{H}^2$-isometry fixing these two ends. By inspection of the standard forms r_θ, t_α, d_ρ of orientation-preserving isometries (Section 4.5), we see that the only one which fixes the ends of a line is d_ρ, which fixes the ends of its axis. Thus, we must have

$$f = \bar{f}\bar{r}_L = \text{translation with axis } L,$$

and therefore

$$\bar{f} = f\bar{r}_L = \text{glide reflection with axis } L. \qquad \square$$

Exercises

4.6.1. Show that if L, M are asymptotic $\mathbb{H}^2$-lines and if L' is any $\mathbb{H}^2$-line sharing the common end of L, M, then there is an $\mathbb{H}^2$-line M' such that $\bar{r}_M\bar{r}_L = \bar{r}_{M'}\bar{r}_{L'}$.

4.6.2. Prove a result analogous to Exercise 4.6.1, where L, M, L' are $\mathbb{H}^2$-lines with a common perpendicular.

4.6.3. Deduce from Exercises 4.6.1 and 4.6.2 that if L, M, N are $\mathbb{H}^2$-lines with a common point (either in $\mathbb{H}^2$ or on the circle at infinity), or a common perpendicular, then $\bar{r}_N\bar{r}_M\bar{r}_L$ is an $\mathbb{H}^2$-reflection.

4.6.4. If M, N have a common point which they do *not* share with L, find M', N' such that $\bar{r}_N\bar{r}_M = \bar{r}_{N'}\bar{r}_{M'}$ and M' is perpendicular to L.

4.6.5. If L, M have a common perpendicular which they do not share with N, find M', L' such that $\bar{r}_M\bar{r}_L = \bar{r}_{M'}\bar{r}_{L'}$ and M' has a common point with N.

4.6.6. Deduce from Exercises 4.6.3, 4.6.4, and 4.6.5 that $\bar{r}_N\bar{r}_M\bar{r}_L$ is an $\mathbb{H}^2$-glide reflection for any $\mathbb{H}^2$-lines L, M, N.

4.7   The Area of a Triangle

The existence of asymptotic lines in the hyperbolic plane gives rise to *asymptotic triangles*—triangles in which two sides are asymptotic. Every ordinary triangle is, in fact, the difference of two asymptotic triangles (see Figure 4.23); hence, we can find the area of an ordinary triangle from that of an asymptotic triangle if the latter is finite—and it is! We calculate area by passing to the $\mathbb{H}^2$-model, where an infinitesimal rectangle of euclidean width dx and euclidean height dy has $\mathbb{H}^2$-width dx/y and $\mathbb{H}^2$-height dy/y; hence, $\mathbb{H}^2$-area

$$dA = dx\, dy/y^2.$$

Thus, to find the area of a triangle Δ we have to integrate $1/y^2$ over Δ. As usual, the trick is to choose Δ in a convenient position.

Theorem. *The area of an asymptotic triangle $\Delta_{\alpha\beta}$ with angles α , β (and $\gamma = 0$) is $\pi - (\alpha + \beta)$.*

Proof. If we first suppose that $\Delta_{\alpha\beta}$ is in $\mathbb{D}^2$, we can rotate about O to arrange that the limit vertex of $\Delta_{\alpha\beta}$ is i. Then the image of $\Delta_{\alpha\beta}$ in $\mathbb{H}^2$ has vertical lines for its asymptotic sides and the third side is a semicircle with center on the real axis. By an $\mathbb{H}^2$-isometry $z \mapsto \delta + z$, for some $\delta \in \mathbb{R}$, we can bring the center of this semicircle to O, and by a further $\mathbb{H}^2$-isometry $z \mapsto \rho z$, for some $\rho \in \mathbb{R}$, we can make the radius 1. Thus, there is no loss of generality in assuming that $\Delta_{\alpha\beta}$ is of the form shown in Fig. 4.22.

It follows from the fact that radii are orthogonal to the circle that the angles at O are α and β as shown. Then, since the length of the radius is 1, we have $\lambda = \cos(\pi - \alpha)$ and $\mu = \cos \beta$. The formula for $\mathbb{H}^2$-area dA gives

$$\text{area}(\Delta_{\alpha\beta}) = \int\int_{\Delta_{\alpha\beta}} \frac{dx\, dy}{y^2} = \int_\lambda^\mu dx \int_{\sqrt{1-x^2}}^\infty \frac{dy}{y^2} = \int_\lambda^\mu \frac{dx}{\sqrt{1-x^2}},$$

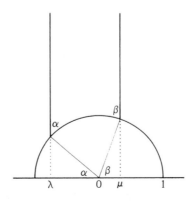

FIGURE 4.22.

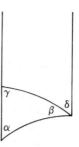

FIGURE 4.23.

and making the change of variable $x = \cos\theta$, we get

$$\int_{\lambda=\cos(\pi-\alpha)}^{\mu=\cos\beta} \frac{dx}{\sqrt{1-x^2}} = \int_{\pi-\alpha}^{\beta} \frac{-\sin\theta\,d\theta}{\sin\theta} = \pi - \alpha - \beta. \qquad \square$$

Corollary. *The area of a triangle $\Delta_{\alpha\beta\gamma}$ with angles $\alpha, \beta, \gamma \neq 0$ is $\pi - (\alpha + \beta + \gamma)$.*

Proof. By prolonging one side of $\Delta_{\alpha\beta\gamma}$ to infinity, we can represent $\Delta_{\alpha\beta\gamma}$ as the difference of two asymptotic triangles $\Delta_{\alpha,\beta+\delta}$ and $\Delta_{\pi-\gamma,\delta}$ as shown in Figure 4.23. Since integrals, and hence areas, are additive we have

$$\begin{aligned}
\text{area}(\Delta_{\alpha\beta\gamma}) &= \text{area}(\Delta_{\alpha,\beta+\delta}) - \text{area}(\Delta_{\pi-\gamma,\delta}) \\
&= \pi - (\alpha + \beta + \delta) - \pi + (\pi - \gamma + \delta) \quad \text{by the theorem} \\
&= \pi - (\alpha + \beta + \gamma). \qquad \square
\end{aligned}$$

Exercises

4.7.1. Show that the formula $\text{area}(\Delta_{\alpha\beta\gamma}) = \pi - (\alpha + \beta + \gamma)$ is also valid when two or more of the angles α, β, γ are zero.

4.7.2. If Σ is any hyperbolic n-gon, show that $\text{area}(\Sigma) = \text{defect}(\Sigma)$, where $\text{defect}(\Sigma) = (n-2)\pi - $ angle sum of Σ. Use the case $n = 2$ to show that the hyperbolic line between two points is unique.

4.7.3. Show that there is an equilateral triangle with angles π/m, for any integer $m \geq 4$. What is the corresponding result for regular n-gons?

4.7.4. Show that any triangle with all angles zero may be mapped to any other by an $\mathbb{H}^2$-isometry.

4.7.5. Show that there is a hyperbolic triangle with angles π/p, π/q, π/r for positive integers p, q, r if and only if $\frac{1}{p} + \frac{1}{q} + \frac{1}{r} < 1$. and that the one with $p, q, r = 2, 3, 7$ has the smallest area.

4.8   The Projective Disc Model

The projective model is the result of mapping the hyperbolic plane into $\mathbb{R}^2$ so that the images of hyperbolic lines are straight. A remarkably simple construction of this model from the conformal disc model, due to Beltrami [1868'], is shown in Figures 4.24 and 4.25. Figure 4.24 shows the disc $\mathbb{D}^2$, with a typical $\mathbb{D}^2$-line, being mapped onto the lower half of a sphere by inverse stereographic projection. We know from Chapter 3 (Exercise 3.3.1 and Section 3.4) that stereographic projection preserves circles and angles; hence, $\mathbb{D}^2$-lines are mapped to circles orthogonal to the equator. It follows that the image of a $\mathbb{D}^2$-line on the lower hemisphere is a vertical section, hence if we project the hemisphere vertically onto the plane $\mathbb{R}^2$, the image of a $\mathbb{D}^2$-line will be straight.

Since this transformation T of $\mathbb{D}^2$ is obviously bijective, it defines a new model of the hyperbolic plane with

Domain, $\mathbb{P}^2 = T\mathbb{D}^2 =$ open unit disc.

$\mathbb{P}^2$-lines $= T$-images of $\mathbb{D}^2$-lines $=$ euclidean line segments across $\mathbb{P}^2$.

$\mathbb{P}^2$-isometries $=$ mappings TgT^{-1}, where g is a $\mathbb{D}^2$-isometry.

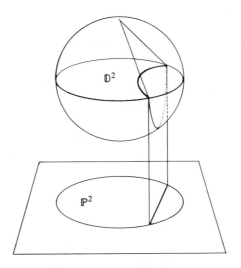

FIGURES 4.24 and 4.25.

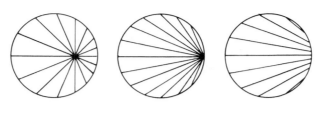

FIGURE 4.26.

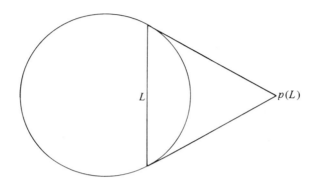

FIGURE 4.27.

This model is called the *projective disc model* because the $\mathbb{P}^2$-isometries are, in fact, projective mappings of $\mathbb{R}^2$, as we shall see below.

The projective model seems less intuitive than $\mathbb{D}^2$ or $\mathbb{H}^2$ because it distorts angles as well as lengths. However, the straightness of $\mathbb{P}^2$-lines has some advantages. It becomes immediately clear that there is a unique hyperbolic line through any two points, for example. Another result which leaps to the eye is: *a hyperbolic polygon is convex if and only if all its angles are* $< \pi$ (Poincaré [1880]). This is certainly true of the euclidean polygons formed by $\mathbb{P}^2$-lines; hence, it is also true of their preimages in $\mathbb{D}^2$ (where angle has its euclidean meaning) because T preserves the property of an angle being $< \pi$.

The projective model also alerts us to a fact which is scarcely discernible in the other models: there is a meaning to points "beyond infinity". We have already seen that a limit rotation can be viewed as a rotation about a point at infinity. We shall now see that a hyperbolic translation can be viewed as a rotation about a point *beyond* infinity. In fact, the families of lines permuted by a rotation, a limit rotation and a translation appear as follows in the projective model (Figure 4.26). Recall from Section 4.5 that the lines permuted by a translation are the perpendiculars to its invariant line. The reason they now have a common point is that the $\mathbb{P}^2$-perpendiculars to any $\mathbb{P}^2$-line L pass through the *pole* $p(L)$ of L, which is the intersection of the

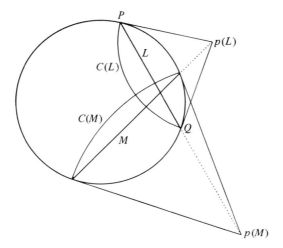

FIGURE 4.28.

tangents to the unit circle through the ends of L (Figure 4.27). This is a consequence of the following:

Orthogonality condition. $\mathbb{P}^2$-*lines L and M are $\mathbb{P}^2$-orthogonal if and only if $p(M)$ lies on L and $p(L)$ lies on M.*

Proof. Let $C(L)$, $C(M)$ be the $\mathbb{D}^2$-lines which T maps onto L, M, respectively (Figure 4.28). Since $C(L)$, $C(M)$ are circles orthogonal to the unit circle, their centers are at the poles $p(L)$, $p(M)$ of L, M, respectively.

Now suppose that $C(L)$ is orthogonal to $C(M)$. Then inversion in $C(M)$ maps both $C(L)$ and the unit circle onto themselves [since both have their intersections and orthogonality with $C(M)$ preserved], and, hence, exchanges their intersections, P and Q. But it is clear from the definition of inversion that if two points are exchanged by an inversion, then the euclidean line through them (in this case L) passes through the center of the circle of inversion [in this case $p(M)$].

Thus, $p(M)$ lies on L, and, hence, $p(L)$ lies on M, by symmetry.

Conversely, if $p(M)$ lies on L, then inversion in $C(M)$ maps L and the unit circle onto themselves, hence exchanges P and Q, hence maps $C(L)$ onto itself [since it maps $C(L)$ onto a circle through P, Q orthogonal to the unit circle, and $C(L)$ is the only such]. But this implies $C(L)$ is orthogonal to $C(M)$. □

The orthogonality condition can be used to prove the seemingly obvious but subtle fact that $\mathbb{P}^2$-isometries extend to projective mappings of $\mathbb{R}^2$, i.e., to mappings which map euclidean lines onto euclidean lines (see Exercises). Thus $\mathbb{P}^2$-isometries are projective mappings which map the disc $\mathbb{P}^2$ onto itself. In fact, $\mathbb{P}^2$-isometries are *all* such mappings, though proving this

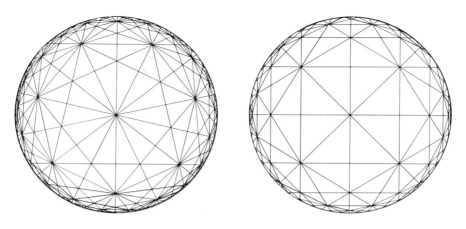

FIGURE 4.29.

requires the so-called *fundamental theorem of projective geometry*, which would take us too far afield (see, e.g., Berger [1987, §4.4]). We merely point out one interesting consequence of this fact: *the only mappings of the hyperbolic plane which map hyperbolic lines onto hyperbolic lines are isometries.*

Pictures of $\mathbb{P}^2$ are seldom seen, perhaps because details are squeezed much closer to the boundary than they are in the $\mathbb{H}^2$- and $\mathbb{D}^2$-models. Still, quite a lot of detail can be seen in tessellations by triangles of small area. Figure 4.29 shows the tessellations by the triangle with angles $\pi/2$, $\pi/3$, $\pi/7$ (left) and by the triangle with angles $\pi/2$, $\pi/4$, $\pi/5$ (right).

Exercises

4.8.1. Show that any two nonintersecting, nonasymptotic hyperbolic lines have a unique common perpendicular.

4.8.2. Call a maximal set of $\mathbb{P}^2$-lines with a common point in $\mathbb{P}^2$ an *internal pencil*, a maximal set of $\mathbb{P}^2$-lines with a common point on the unit circle $\partial\mathbb{P}^2$, a *boundary* pencil, and a maximal set of $\mathbb{P}^2$-lines with a common point not in $\mathbb{P}^2 \cup \partial\mathbb{P}^2$ an *external* pencil. Show that a $\mathbb{P}^2$-isometry sends each pencil to a pencil of the same type.

4.8.3. Extend a $\mathbb{P}^2$-isometry $g : \mathbb{P}^2 \to \mathbb{P}^2$ to a map $g^* : \mathbb{R}^2 \to \mathbb{R}^2$ by letting the g^*-image of $P \notin \mathbb{P}^2$ be the center of the g-image pencil of the pencil through P. Show that g^* maps euclidean lines through $\mathbb{P}^2$ onto euclidean lines through $\mathbb{P}^2$.

4.8.4. Show that the set of poles of members of an internal or boundary pencil forms a euclidean line outside $\mathbb{P}^2$.

4.8.5. Deduce that g^* maps euclidean lines outside $\mathbb{P}^2$ onto euclidean lines outside $\mathbb{P}^2$.

4.9 Hyperbolic Space

The three-dimensional analogue of $\mathbb{H}^2$ is the *half-space model* $\mathbb{H}^3$ of *hyperbolic space*. Its domain is the upper half-space $\{(x, y, z) \mid z > 0; x, y, z \in \mathbb{R}\}$ in (x, y, z)-space $\mathbb{R}^3$, and the $\mathbb{H}^3$-isometries are all products of inversions in spheres orthogonal to the (x, y)-plane. Inversion in a sphere is defined analogously to inversion in a circle, and one can calculate from the definition that

$$\frac{ds}{z} = \frac{\sqrt{dx^2 + dy^2 + dz^2}}{z}$$

is invariant under these inversions (Exercise 4.9.1). Hence, if we take $\int_K ds/z$ as the $\mathbb{H}^3$-*length* of a curve K, we get a distance function on $\mathbb{H}^3$ under which the inversions are isometries.

The $\mathbb{H}^3$-*planes* are hemispheres orthogonal to the (x, y)-plane, which (together with the point ∞) is the *sphere at infinity* of $\mathbb{H}^3$, or else vertical half-planes (which are regarded as hemispheres of infinite radius). $\mathbb{H}^3$-*lines* are intersections of $\mathbb{H}^3$-planes, i.e., semicircles orthogonal to the sphere at infinity (Exercise 4.9.2). Thus, the geometry of $\mathbb{H}^3$ can be developed quite analogously to the geometry of $\mathbb{H}^2$. We do not wish to do this because three-dimensional hyperbolic geometry is beyond our scope; however, $\mathbb{H}^3$ is also remarkable as a showcase of two-dimensional geometry. As we shall now show, $\mathbb{H}^3$ contains not only different models of the hyperbolic plane, but also euclidean planes and spheres!

First, where are the models of the hyperbolic plane? Naturally, they should include the $\mathbb{H}^3$-planes. Since $\mathbb{H}^3$-distance is ds/z, i.e., euclidean distance divided by the euclidean height above the (x, y)-plane, any vertical half-plane is an $\mathbb{H}^2$. The ordinary hemispheres orthogonal to the (x, y)-plane, on which the $\mathbb{H}^3$-lines are vertical semicircles, turn out to be identical with the *hemisphere model* constructed in Section 4.8 as an intermediate stage between $\mathbb{D}^2$ and $\mathbb{P}^2$. Figure 4.30 shows an $\mathbb{H}^2$ and a hemisphere model in $\mathbb{H}^3$.

$\mathbb{P}^2$ itself appears in a remarkable way as the "distant view" of an $\mathbb{H}^3$-plane, if one assumes that light rays in $\mathbb{H}^3$ travel along $\mathbb{H}^3$-lines. Since a hyperbolic object increases its euclidean size as z increases, the eye of a creature in $\mathbb{H}^3$ can completely cover an $\mathbb{H}^3$-plane when moved to a position with a sufficiently large z-value (Figure 4.31). Since, also, the $\mathbb{H}^3$-lines include the vertical euclidean lines, the $\mathbb{H}^3$-creature will see the projection of the hemisphere model on the (x, y)-plane, i.e., the $\mathbb{P}^2$-model! The circle at infinity of this $\mathbb{P}^2$-model is the visual horizon of this $\mathbb{H}^3$-plane. (Look again at Figure 4.29 and see whether you can get the sensation of looking down on a hemisphere.)

The euclidean spheres tangential to the boundary of $\mathbb{H}^3$ are the *horospheres*, the analogues of horocycles in $\mathbb{H}^2$. These include the horizontal euclidean planes $z = $ constant, which can be regarded as spheres tangential to the point ∞ of the (x, y)-plane. On the horosphere $z = $ constant, we

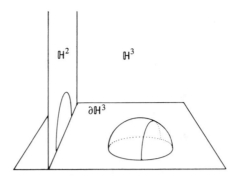

FIGURE 4.30.

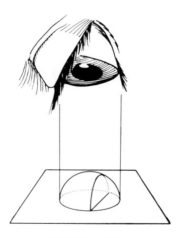

FIGURE 4.31.

have $\mathbb{H}^3$-distance = (euclidean distance)/constant; hence, the geometry is precisely that of the euclidean plane $\mathbb{R}^2$. Moreover, all the euclidean isometries can be induced by $\mathbb{H}^3$-isometries because one can induce any reflection of $\mathbb{R}^2$ by a reflection in an $\mathbb{H}^3$-plane.

Finally, it can be shown that the geometry of an $\mathbb{H}^3$-*sphere*—the set of points at constant $\mathbb{H}^3$-distance from a point of $\mathbb{H}^3$—is the same as that of $\mathbb{S}^2$. And, not surprisingly, the isometries of $\mathbb{S}^2$ can be induced by isometries of $\mathbb{H}^3$. Thus, $\mathbb{H}^3$ really is an extraordinarily accommodating space, far more so than $\mathbb{R}^3$ which, as we have said, will not accommodate any reasonable $\mathbb{H}^2$ at all, let alone provide isometries for it.

As if this were not enough, $\mathbb{H}^3$ also models the *geometry of all linear fractional transformations of* $\mathbb{C} \cup \{\infty\}$. We interpret the (x, y)-plane as the plane of a complex variable $w = x + iy$, and observe that any linear fractional function $w \mapsto \frac{aw+b}{cw+d}$ with $ad - bc \neq 0$ is a product of inversions because

$$\frac{aw+b}{cw+d} = \frac{\frac{a}{c}(cw+d)+b-\frac{ad}{c}}{cw+d} = \frac{a}{c} + \frac{bc-ad}{c(cw+d)}.$$

This equation shows that $w \mapsto \frac{aw+b}{cw+d}$ is the product of functions of the forms

$$w \mapsto pw, \quad w \mapsto q+w, \quad w \mapsto 1/w \quad \text{where } p, q \in \mathbb{C},$$

each of which is easily shown to be a product of inversions (Exercise 4.9.3). For good measure, we can also express functions $w \mapsto \frac{a\bar{w}+b}{c\bar{w}+d}$ by composing with the inversion $w \mapsto \bar{w}$.

Now inversion in a circle in the (x, y)-plane extends to inversion in a hemisphere orthogonal to the (x, y)-plane, i.e., to an $\mathbb{H}^3$-isometry. Thus, *any linear fractional function of w or $\bar{w}$ can be realized as the mapping induced on the sphere at infinity by an isometry of $\mathbb{H}^3$.* Conversely, *any isometry of $\mathbb{H}^3$ induces a linear fractional mapping of the sphere at infinity* since, by definition, $\mathbb{H}^3$-isometries induce mappings of the sphere at infinity which are products of inversions, and, hence, linear fractional by Section 3.4.

All at once the various representations of euclidean, spherical, and hyperbolic isometries by linear fractional functions are seen to be parts of the one big picture. Since euclidean, spherical and hyperbolic isometries can all be accommodated within $\mathbb{H}^3$, they can all be reflected in the induced mappings of the sphere at infinity.

Exercises

4.9.1. Show that inversion in the unit sphere $\mathbb{S}^2$ in $\mathbb{R}^3$ is

$$(x, y, z) \mapsto \left(\frac{x}{x^2+y^2+z^2}, \frac{y}{x^2+y^2+z^2}, \frac{z}{x^2+y^2+z^2} \right).$$

Deduce that inversion in $\mathbb{S}^2$, and hence in any sphere with center on the (x, y)-plane, preserves ds/z.

4.9.2. Show that the points common to the spheres

$$(x-a_1)^2 + (y-b_1)^2 + z^2 = r_1^2 \quad \text{and} \quad (x-a_2)^2 + (y-b_2)^2 + z^2 = r_2^2$$

lie in a vertical plane.

4.9.3. Show that the mappings $w \mapsto pw$, $w \mapsto q+w$ and $w \mapsto 1/w$ are products of inversions.

4.9.4 (l'Hopital, Jean Bernoulli). If the half-space $z > 0$ is filled by a medium with the property that the speed of light in it is proportional to z, show that light travels along $\mathbb{H}^3$-lines. (Hint: Choose a metric on $z > 0$ for which the speed of light is constant, and assume Fermat's principle that light takes least time.)

4.10   Discussion

The discovery of a hyperbolic plane, modeling the non-euclidean geometry of Gauss, Bolyai and Lobachevsky, is due to Beltrami [1868]. Following up his discovery [1865] that the surfaces which can be mapped into $\mathbb{R}^2$ in such a way that their geodesics go to lines are precisely those of constant curvature, he constructed a mapping of the pseudosphere into $\mathbb{R}^2$ which led him to the projective model $\mathbb{P}^2$. His [1868'] expands these ideas to encompass the conformal models as well, and in all dimensions. The discovery that hyperbolic geometry is the geometry of linear fractional transformations is due to Poincaré. He interpreted the real transformations $z \mapsto \frac{\alpha z + \beta}{\gamma z + \delta}$ as isometries of $\mathbb{H}^2$ in his [1882], and the general transformations $z \mapsto \frac{az+b}{cz+d}$ as isometries of $\mathbb{H}^3$ in his [1883]. This, in fact, showed that real transformations have a double geometric meaning because they are also the conformal mappings of $\mathbb{H}^2$ (see Jones and Singerman [1987, p. 200]).

Even more remarkable, the *points* of $\mathbb{H}^2$ can themselves be interpreted as geometric objects—the "parallelogram shapes" mentioned in Section 2.10 in connection with elliptic functions—and this interpretation of $\mathbb{H}^2$ gives rise to the best-known group of linear fractional transformations, the *modular group*, as follows.

The shape of the parallelogram with vertices 0, ω_1, ω_2, $\omega_1 + \omega_2 \in \mathbb{C}$ is determined by the complex number ω_1/ω_2 because $\arg(\omega_1/\omega_2) = \arg \omega_1 - \arg \omega_2$ is the angle between adjacent sides and $|\omega_1/\omega_2| = |\omega_1|/|\omega_2|$ is the ratio of their lengths. By suitable ordering of the vertices, we can ensure that $\omega_1/\omega_2 = \tau$ is in $\mathbb{H}^2$. Conversely, each τ in $\mathbb{H}^2$ corresponds to a parallelogram shape.

Thus, $\mathbb{H}^2$ is the space of parallelogram shapes. Now we ask: when do two parallelogram shapes determine the same torus shape? (Like parallelograms, tori admit similarities, and we say two tori are "of the same shape" when there is a similarity mapping one onto the other. It so happens that similarities are the only conformal mappings between tori—see Jones and Singerman [1987, p. 202]—so the shape of a torus captures its geometric essence.) Just as more than one parallelogram determines the same torus, more than one parallelogram shape determines the same torus shape. For example, all the parallelograms shown in Figure 4.32 determine the same torus shape, the "square" torus.

The problem is to decide when two pairs $\{\omega_1, \omega_2\}$ and $\{\omega_1', \omega_2'\}$ generate the same subgroup of $\mathbb{C}$. The answer was found by Lagrange [1773] in his study of quadratic forms. In our notation it is

$$\tau' = \frac{\alpha\tau + \beta}{\gamma\tau + \delta},$$

where $\tau = \omega_1/\omega_2$, $\tau' = \omega_1'/\omega_2'$ and α, β, γ, δ are integers with $\alpha\beta - \gamma\delta = \pm 1$. Thus, the various parallelogram shapes which determine the

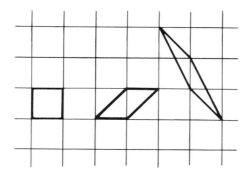

FIGURE 4.32.

same torus shape are related by isometries of $\mathbb{H}^2$ with integer coefficients and determinant ± 1. The group of these isometries is called the (extended) modular group because $\tau = \omega_1/\omega_2$ was originally called the *modulus* (back in the 18th century when it was a mere coefficient in the denominator of an elliptic integral). The orbit of τ corresponds to the shape of the torus obtained by pasting opposite sides of the parallelogram with shape τ. Taking one leap further, we can view the quotient $\mathbb{H}^2/(\text{modular group})$ as *moduli space*, the space of "shapes of tori".

Actually this is getting ahead of ourselves because $\mathbb{H}^2/(\text{modular group})$ is not a true surface, but a more general structure called an *orbifold* (see Chapter 8). Still, it gives us a glimpse of the deep and varied connections between euclidean and hyperbolic geometry. We shall be in a better position to understand quotients such as $\mathbb{H}^2/(\text{modular group})$ after we have studied quotients of $\mathbb{H}^2$ by simpler groups (Chapter 5). Readers may also enjoy the elementary geometric approach to tori and the modular group in Nikulin and Shafarevich [1987].

5

Hyperbolic Surfaces

5.1 Hyperbolic Surfaces and the Killing–Hopf Theorem

The concept of hyperbolic surface is completely analogous to the concept of euclidean surface (Section 2.7). A *hyperbolic surface* is a set S with a real-valued distance function d_S such that each $P \in S$ has an ϵ-neighborhood isometric to a disc of $\mathbb{H}^2$. The proof of the Killing–Hopf theorem (Section 2.9) carries over word-for-word (provided "line", "distance" etc., are understood in the hyperbolic sense), showing that any complete, connected hyperbolic surface is of the form $\mathbb{H}^2/\Gamma$, where Γ is a discontinuous, fixed point free group of $\mathbb{H}^2$-isometries.

As in the euclidean case, the fact that Γ is fixed point free means that it cannot contain rotations. The only other possibilities in the euclidean case are translations and glide reflections, but in the hyperbolic case there is a third possibility: Γ can contain limit rotations.

It is at this point that the hyperbolic theory begins to diverge from the euclidean. Limit rotations in Γ yield a quotient surface $\mathbb{H}^2/\Gamma$ with one or more *cusps*—infinitely long spikes whose diameter tends to zero. We shall study some surfaces with cusps in Sections 5.2 to 5.4. Even if Γ contains only translations, there is no longer any bound on the number of generators of Γ. (Contrast with Section 2.5.) This leads to quotient surfaces of infinitely many different topological types—like the torus but with more "holes"—which we shall look at in Section 5.6.

The infinite multiplicity of possibilities for Γ in the hyperbolic case rather changes the problem of finding the possible quotients $\mathbb{H}^2/\Gamma$ (which, by the Killing–Hopf theorem, are all the complete hyperbolic surfaces). We cannot simply enumerate finitely many types of Γ as in the euclidean case (Section 2.5). In fact, it proves to be no easier to enumerate the groups Γ than to find the surfaces $\mathbb{H}^2/\Gamma$ directly. The group theory and geometry meet on the common ground of *polygons*, which appear, on the one hand, as fundamental regions for groups, and on the other as "raw material" for the construction of surfaces, by "pasting" their edges.

The interaction between group theory and geometry in $\mathbb{H}^2$ via polygons is really the main theme of this chapter. Polygons are easier to understand than either groups or hyperbolic surfaces, so it is not surprising that the classification of both the groups Γ and the surfaces $\mathbb{H}^2/\Gamma$ reduces to the classification of polygons. We shall see how this comes about in Section 5.5.

5.2   The Pseudosphere

The classical pseudosphere (Section 4.1) is an incomplete surface, but it has a natural completion $\mathbb{H}^2/\Gamma$, where $\Gamma = \langle t_{2\pi} \rangle$ is the group of limit rotations of $\mathbb{H}^2$ generated by $t_{2\pi}(z) = 2\pi + z$. The complete pseudosphere is easily understood in terms of a tessellation of $\mathbb{H}^2$ by vertical strips (fundamental regions for Γ), any one of which becomes $\mathbb{H}^2/\Gamma$ when its opposite sides are identified by $t_{2\pi}$ (Figure 5.1, which also shows a typical Γ-orbit). This also shows that the pseudosphere is homeomorphic to a cylinder. The "end" where $y \to \infty$ is called the *cusp* because its $\mathbb{H}^2$-width $\to 0$ as its $\mathbb{H}^2$-height $\to \infty$.

The *lines* on $\mathbb{H}^2/\Gamma$ are defined, as usual, to be the Γ-images of $\mathbb{H}^2$-lines. We can visualize the course of the lines on the pseudosphere by superimposing $\mathbb{H}^2$-lines on the tessellation of $\mathbb{H}^2$ by fundamental regions for Γ. Each time a line crosses the edge of a fundamental region R, one imagines it reentering R at the equivalent point on the opposite edge (corresponding to a crossing of the "join" on the pseudosphere). Thus, to find the image ΓL on $\mathbb{H}^2/\Gamma$ of a line $L \subset \mathbb{H}^2$, one draws all images gL of L, for $g \in \Gamma$, which intersect a given fundamental region R. For example, if L is as shown in Figure 5.2, then its translates by 1, $t_{2\pi}$ and $t_{2\pi}^2$ all meet the shaded fundamental region R. This shows, in particular, that the image of L on $\mathbb{H}^2/\Gamma$ crosses itself twice.

The only $\mathbb{H}^2$-lines which go to infinity in the y-direction are the lines $x = $ constant, corresponding to tractrix sections of the classical pseudosphere. It follows that the only lines on the pseudosphere which go all the way along the cusp are the tractrix sections. Any line which winds around the surface goes only a finite distance in the direction of the cusp before turning back.

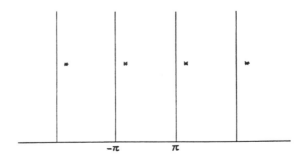

FIGURE 5.1.

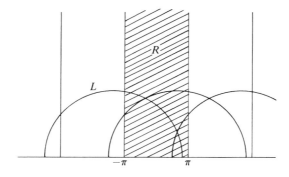

FIGURE 5.2.

There are no "helical" lines as there are on the euclidean cylinder.

Exercises

5.2.1. Show that each line on the pseudosphere intersects itself, or another line, only finitely often.

5.2.2 (Huygens). Show that the classical pseudosphere has finite area.

5.2.3. Show that any surface $\mathbb{H}^2/\Gamma$, where Γ is generated by a limit rotation $g(z) = \alpha + z$, is isometric to the pseudosphere.

5.2.4. Show that the complete pseudosphere can be embedded in the hyperbolic space $\mathbb{H}^3$ of Section 4.9.

5.3   The Punctured Sphere

In Section 2.7 we poured cold water on the idea of constructing a surface by removing points from the plane, pointing out that the resulting surface would be *incomplete* because not all lines could be continued indefinitely. However, this incompleteness was with respect to the euclidean distance function, and we now know that other distance functions are possible. Indeed, we have already seen that the euclidean plane can be viewed as a sphere minus one point—by stereographic projection from the missing point—so a sphere which is incomplete with respect to spherical distance is actually complete with respect to euclidean distance. Of course, in calling the plane a "sphere minus one point," we mean only that it is *homeomorphic* to a standard sphere minus one point. Since we are considering replacements for the spherical distance function, there is no need for "sphere" to mean more than this.

The sphere minus one point can also be made complete with respect to

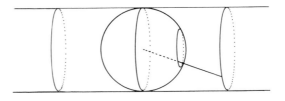

FIGURE 5.3.

hyperbolic distance, e.g., by mapping it homeomorphically onto $\mathbb{D}^2$ (Exercise 5.3.1) and defining the distance between two points on the sphere to be the $\mathbb{D}^2$-distance between their images.

A little thought shows that the sphere minus two points, which is homeomorphic to the plane minus one point, can also be made complete with respect to euclidean or hyperbolic distance. The sphere minus two points is homeomorphic to a cylinder, and an explicit homeomorphism is the central projection (Figure 5.3). The cylinder has an obvious euclidean distance function (Section 2.2), and it also has a hyperbolic one (which makes it a pseudosphere, Section 5.2). The sphere minus three points *cannot* be given a euclidean distance function because none of the euclidean surfaces (Section 2.9) is homeomorphic to it. However, it can be given a hyperbolic distance function, as can a sphere with any finite number of missing points ("punctures"). The rest of this section is devoted to the hyperbolic sphere with three punctures; the case of more punctures is similar but more complicated (see Exercise 5.5.1).

Consider the region $R \subset \mathbb{H}^2$ bounded by $\mathrm{Re}(z) = \pm 1$ and the semicircles $|z \pm \frac{1}{2}| = \frac{1}{2}$ (Figure 5.4). It is clear that we can make R into a topological sphere with three punctures by identifying the left boundary $\mathrm{Re}(z) = -1$ with the right boundary $\mathrm{Re}(z) = 1$, and the lower left boundary $|z + \frac{1}{2}| = \frac{1}{2}$ with the lower right, $|z - \frac{1}{2}| = \frac{1}{2}$, in the manner indicated by Figure 5.5. The reason we are able to obtain a hyperbolic distance function on this punctured sphere is that the identifications can be realized by hyperbolic isometries $g(z) = 2 + z$ and $h(z) = \frac{z}{2z+1}$. [It is obvious that g does the right thing; to see that h also works, express $|z + \frac{1}{2}| = \frac{1}{2}$ in the parametric form $z = \frac{1}{2}(e^{i\theta} - 1)$. Then h sends z to $\frac{1}{2}(e^{i\theta} - 1)/e^{i\theta} = \frac{1}{2}(1 - e^{-i\theta})$, which is a parametric form of $|z - \frac{1}{2}| = \frac{1}{2}$.]

The points of the punctured sphere are the points in the interior of R and pairs of identified points $\{z, gz\}$ or $\{z, hz\}$ on the boundary of R. Each z in the interior has a hyperbolic disc $D \subset R$ as its neighborhood disc. A z on the boundary lies in two discs, one the image of the other under either g or h, depending on the part of the boundary on which it lies (Figure 5.5).

This is enough to guarantee that the result of identifying the sides of R is a hyperbolic surface, as we shall see in Section 5.5. However, it is not clear that the surface is complete, which is what we need to apply the Killing–

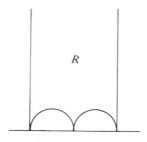

FIGURE 5.4.

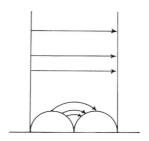

FIGURE 5.5.

Hopf theorem and conclude that the punctured sphere is $\mathbb{H}^2/\Gamma$, where Γ is the group $\langle g, h \rangle$ generated by g and h. Instead, we shall prove directly that R is a fundamental region for Γ.

Theorem. *R is a fundamental region for $\Gamma = \langle g, h \rangle$.*

Proof. We have to show that the images wR of R, for $w \in \langle g, h \rangle$, meet at most along edges and fill $\mathbb{H}^2$. Calculation shows that $g^{\pm 1}R$, $h^{\pm 1}R$ meet R as shown in Figure 5.6(i). The image of this picture under w, shown schematically in Figure 5.6(ii), reveals that, in general, wR has neighbors $wg^{\pm 1}R$, $wh^{\pm 1}R$, which it meets along edges.

Figure 5.6(ii) shows that, for any distinct $l, l' \in \{g, g^{-1}, h, h^{-1}\}$, wR separates wlR from $wl'R$. If R_1, R_2, R_3 are any regions wR, we will write

$$R_1 \mid R_2 \mid R_3$$

to denote that R_1 is separated, by R_2, from R_3. It follows, e.g., that if $R_1 \mid R_2 \mid R_3$ and $R_2 \mid R_3 \mid R_4$, then $R_1 \mid R_2 \mid R_4$ (and also $R_1 \mid R_3 \mid R_4$). In terms of this notation, the separations exhibited by Figure 5.6(ii) can be written

$$WlR \mid wR \mid wl'R \tag{1}$$

for any distinct $l, l' \in \{g, g^{-1}, h, h^{-1}\}$.

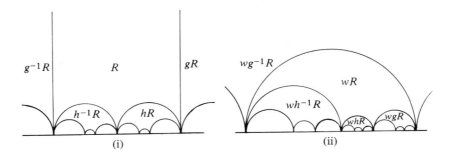

FIGURE 5.6.

We now prove that if $w = l_k \ldots l_1$ is a product of at least two terms $g^{\pm 1}$, $h^{\pm 1}$, with no adjacent inverses (such a product is called *reduced*), then wR is disjoint from R.

We consider the region $l_k \ldots l_1 R = l_k \ldots l_2 \cdot l_1 R$. Since $l_2^{-1} \neq l_1$, by the assumption that w is reduced, we can conclude from (1) that

$$l_k \ldots l_2 \cdot l_1 R \mid l_k \ldots l_2 R \mid l_k \ldots l_2 \cdot l_2^{-1} R,$$

i.e.,

$$l_k \ldots l_2 l_1 R \mid l_k \ldots l_2 R \mid l_k \ldots l_3 R.$$

One similarly proves $l_k \ldots l_2 R \mid l_k \ldots l_3 R \mid l_k \ldots l_4 R \ldots$ and finally $l_k l_{k-1} R \mid l_k R \mid R$.

Combining these results gives

$$l_k \ldots l_1 R \mid l_k \ldots l_2 R \mid R,$$

and hence $wR = l_k \ldots l_1 R$ is disjoint from R.

We already know that wR meets R only along edges if $w \in \{g, g^{-1}, h, h^{-1}\}$; hence wR does not overlap R for any $w \neq 1$. The same is true for $w_1 R$ and $w_2 R$ when $w_1 \neq w_2$ because $w_1 R$, $w_2 R$ are the images of $w_2^{-1} w_1 R$ and R under w_2.

To show that the regions wR fill $\mathbb{H}^2$, we first observe that their union is $\mathbb{H}^2$-convex, i.e., if the union contains points P, Q it also contains the $\mathbb{H}^2$-line segment PQ. Indeed, if $P \in R$ and $Q \in l_k \ldots l_2 l_1 R$, then the union of the successive adjacent cells $l_k \ldots l_2 l_1 R$, $l_k \ldots l_2 R, \ldots, R$ is an $\mathbb{H}^2$-polygon with all vertices on $\partial \mathbb{H}^2$; hence it is $\mathbb{H}^2$-convex and contains PQ. Similarly if $P \in w_1 R$ and $Q \in w_2 R$ because this is the w_1-image of the situation just described.

It now suffices to show that there are regions arbitrarily close to any point of $\partial \mathbb{H}^2$. This will follow if we can show that the vertices of regions are dense on $\partial \mathbb{H}^2$ because this implies that each point of $\partial \mathbb{H}^2$ is spanned by arbitrarily short edges of regions. In fact, we shall show that the vertices include all rational points on $\partial \mathbb{H}^2$.

Consider a rational number m/n. This number is mapped as follows by g, g^{-1}, h, h^{-1}:

$$g\left(\frac{m}{n}\right) = \frac{m+2n}{n}, \quad g^{-1}\left(\frac{m}{n}\right) = \frac{m-2n}{n},$$
$$h\left(\frac{m}{n}\right) = \frac{m}{2m+n}, \quad h^{-1}\left(\frac{m}{n}\right) = \frac{m}{-2m+n}.$$

Thus, if we represent m/n by the pair (m,n), we are able to form the pairs $(m',n') = (m \pm 2n, n)$, $(m, \pm 2m + n)$ by application of $g^{\pm 1}$, $h^{\pm 1}$. Now if $0 \neq |m| < |n|$, we have either $|2m + n| < |n|$ or $|-2m + n| < |n|$. If $0 \neq |n| < |m|$, either $|m + 2n| < |m|$ or $|m - 2n| < |m|$. Hence, if we have different $|m|$, $|n| > 0$ at least one of the pairs (m',n') satisfies

$$|m'| + |n'| < |m| + |n|.$$

By repeated application of $g^{\pm 1}$, $h^{\pm 1}$, we can therefore obtain either $m' = 0$, $n' = 0$, or $|m'| = |n'|$. In other words, $m'/n' = 0$, ∞ or ± 1. Since 0, ∞, ± 1 are the vertices of R, it follows, by applying the inverse sequence of operations $g^{\pm 1}$, $h^{\pm 1}$, that m/n is a vertex of some wR. □

The proof above has two important corollaries. The first, together with the theorem, shows that $\mathbb{H}^2/\Gamma$ is a surface, and hence equals the punctured sphere obtained by identifying edges of R via g and h.

Corollary 1. $\Gamma = \langle g, h \rangle$ *acts on* $\mathbb{H}^2$ *without fixed points.*

Proof. It will suffice to show that no point of R is fixed by an element $w \neq 1$ of $\langle g, h \rangle$ because $P \in R$ is fixed by w if and only if $vP \in vR$ is fixed by vwv^{-1}. Certainly $w \neq g^{\pm 1}$, $h^{\pm 1}$ moves all points of R because only the regions $g^{\pm 1}R$, $h^{\pm 1}R$ touch R. But $w = g^{\pm 1}$, $h^{\pm 1}$ also move all points of R because $g^{\pm 1}$ fixes only ∞ and $h^{\pm 1}$ fixes only 0. □

The second corollary shows that $\Gamma = \langle g, h \rangle$ is a *free group*, i.e., that no nontrivial reduced word w on $g^{\pm 1}$, $h^{\pm 1}$ equals 1. Γ is perhaps the simplest "naturally occurring" example of a free group. (For a related matrix example see Exercise 5.3.4.)

Corollary 2. *If* w *is a nontrivial reduced word, then* $w \neq 1$.

Proof. It is immediate from the proof of the theorem that $wR \neq R$, hence w is not the identity map. □

Exercises

5.3.1. Find a homeomorphism of the sphere minus one point onto the open disc.

5.3.2. Show that a cusp of the punctured sphere, when cut off by a suitable horocycle, is isometric to the cusp of the pseudosphere.

5.3.3. Show that $g^{-1} = r_\pi h r_\pi^{-1}$, where g, h are as above and $r_\pi : z \mapsto -1/z$ is the hyperbolic rotation through π about i.

5.3.4. Show that $\begin{pmatrix} 1 & 2 \\ 0 & 1 \end{pmatrix}$ and $\begin{pmatrix} 1 & 0 \\ 2 & 1 \end{pmatrix}$ generate a free group under matrix multiplication.

5.3.5. Show that $R = R_g \cap R_h$ where R_g is a certain fundamental region for $\langle g \rangle$ and R_h is a certain fundamental region for $\langle h \rangle$.

5.4   Dense Lines on the Punctured Sphere

In Section 2.6 we showed the existence of lines on the torus which are *dense* in the sense that they pass arbitrarily close to every point. On the punctured sphere there are lines which are even more remarkable. Not only do they pass arbitrarily close to any point, but they travel arbitrarily close to every direction through that point. More precisely: there is a line L on the punctured sphere which, for each finite line segment σ and each $\epsilon > 0$, contains a segment $L_{\sigma, \epsilon}$ lying within hyperbolic distance ϵ of σ over the whole length of σ.

The key to this result is the lemma below, which states that a line on the punctured sphere can make an essentially arbitrary series of windings around the cusps. To state the winding property precisely we "mark" the punctured sphere by the two lines that are the images of the edges of fundamental region R (from Section 5.3) under the orbit map $\mathbb{H}^2 \to \mathbb{H}^2/\Gamma$. Figure 5.7 suggests how these lines might look. The arrows g and h show which line on $\mathbb{H}^2/\Gamma$ is which, but also, more importantly, they enable a line on $\mathbb{H}^2/\Gamma$ to be specified by its *edge crossing sequence*.

Definitions. The *edge crossing sequence* of a directed line L on $\mathbb{H}^2/\Gamma$ (other than the marked lines) is the doubly infinite sequence

$$\ldots c_{-1},\ c_0,\ c_1,\ c_2, \ldots$$

of labels— g, g^{-1}, h or h^{-1}—of the successive marked lines crossed by L. The negative exponent is taken when the marked line is crossed against the direction of the arrow. An edge crossing sequence is *reduced* if it contains no consecutive inverse terms.

Lemma. *Any nonconstant, reduced doubly-infinite sequence* $\ldots c_{-1},\ c_0,\ c_1,$ $c_2, \ldots$ *can be realized as the edge crossing sequence of a directed line on the punctured sphere.*

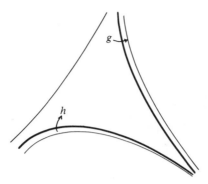

FIGURE 5.7.

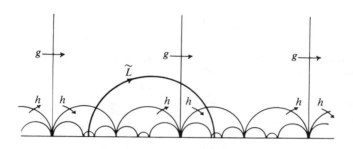

FIGURE 5.8.

Proof. A line L on the punctured sphere $\mathbb{H}^2/\Gamma$ is, by definition, $\Gamma\tilde{L}$ for some line $\tilde{L}$ on $\mathbb{H}^2$. Corresponding to the edge crossings of L on $\mathbb{H}^2/\Gamma$, we have crossings of $\tilde{L}$ from one cell to another in the tessellation of $\mathbb{H}^2$ by the Γ-images of R. The edge crossing sequence of $\tilde{L}$ (and hence of L) is just the sequence of crossings from one cell to the next in the chain of cells through which $\tilde{L}$ passes. Figure 5.8 shows crossings in $\mathbb{H}^2$ labeled by g and h arrows, together with a directed line $\tilde{L}$.

Since each cell has one g arrow in, one out, and one h arrow in, one out, we can realize any reduced edge crossing sequence $\ldots c_{-1}, c_0, c_1, \ldots$ by a chain of distinct cells $\ldots C_{-1}, C_0, C_1, \ldots$ such that the crossing from C_i to C_{i+1} is c_i. We now look at the sequence of edges crossed in the projective model $\mathbb{P}^2$ of Section 4.8 (Figure 5.9). The euclidean length of the line labeled c_i tends to 0 as i tends to $\pm\infty$, by the proof of the theorem in Section 5.3 (ends on $\partial\mathbb{P}^2$ are continuous images of ends on $\partial\mathbb{H}^2$, hence also dense). Consequently, these edges converge to limit points c_∞ and $c_{-\infty}$ on $\partial\mathbb{P}^2$ as i tends to $\pm\infty$. The latter points will be distinct unless the crossing sequence is constant.

Now the union $\ldots C_{-1} \cup C_0 \cup C_1 \cup \ldots$ in $\mathbb{P}^2$ is obviously convex in the

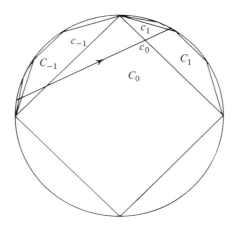

FIGURE 5.9.

euclidean sense, and, hence, also in the hyperbolic sense because $\mathbb{P}^2$-lines are euclidean. In particular, $\ldots C_{-1} \cup C_0 \cup C_1 \cup \ldots$ contains the $\mathbb{P}^2$-line from $c_{-\infty}$ to c_∞. The edge crossing sequence of this line is $\ldots c_{-1}, c_0, c_1, \ldots,$ by construction, and hence the sequence is also realized by the image line on the punctured sphere.   □

Remark. The lemma gives a simple example of a dynamical system with "quasi-random" behavior. If a particle moves freely on a surface of constant curvature, then it travels along a line, and one can ask whether past crossings of marked lines enable future crossings to be predicted. The lemma shows that even if the complete past crossing sequence on the punctured sphere is known, the future crossing sequence is completely unpredictable.

Theorem. *There is a line L on the punctured sphere which, for each finite line segment σ on the punctured sphere and each $\epsilon > 0$, contains a segment $L_{\sigma,\epsilon}$ lying within hyperbolic distance ϵ of σ over the whole length of σ.*

Proof. Given a finite line segment σ on the punctured sphere $\mathbb{H}^2/\Gamma$, we consider an $\mathbb{H}^2$-line segment $\tilde{\sigma}$ whose image is σ under the orbit map. If we prolong $\tilde{\sigma}$ at both ends, passing through a finite chain of cells, we eventually reach cells C_{-n} (in one direction) and C_n (in the other) with the following property: *any* line segment from C_{-n} to C_n passes within hyperbolic distance ϵ of each end of σ. This is easiest to see by viewing the cells in $\mathbb{P}^2$ again, where the euclidean lengths of their edges tend to 0 (Figure 5.10). Hyperbolic perpendiculars of hyperbolic length ϵ have been erected at both ends of $\tilde{\sigma}$, and it is clear that, once C_{-n} and C_n are sufficiently small (in the euclidean sense), any line segment $\tilde{L}_{\sigma,\epsilon}$ from one to the other will meet these ϵ-perpendiculars to $\tilde{\sigma}$ at both ends.

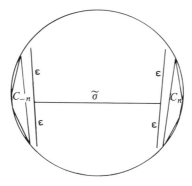

FIGURE 5.10.

This shows how to construct a line $\tilde{L}$ in $\mathbb{H}^2$ containing a segment $\tilde{L}_{\sigma,\epsilon}$ whose image $L_{\sigma,\epsilon}$ has the required property for a *single* pair σ, ϵ. We have only to ensure that $\tilde{L}$ makes a finite sequence of edge crossings identical with the sequence made by a line segment from C_{-n} to C_n. The cells actually crossed do *not* have to be those between C_{-n} and C_n. A line segment $\tilde{L}'_{\sigma,\epsilon}$ with the same image $L_{\sigma,\epsilon}$ as $\tilde{L}_{\sigma,\epsilon}$ can be started in any cell.

But then we can also construct a line L with the required property for *all* pairs σ, ϵ. Enumerate the countably many reduced finite sequences of edge crossings (a rare use of set theory in geometry!), incorporate them in one reduced doubly-infinite sequence, and apply the lemma. $\square$

Remark. A hyperbolic triply punctured sphere can be constructed, in a certain sense, in the hyperbolic space $\mathbb{H}^3$ of Section 4.9. One takes a copy of the fundamental region R as part of a vertical plane in $\mathbb{H}^3$. Then if R is folded down the middle, one can make the required edge identifications (Figure 5.11), except that the surface is "flattened" in the process. Its "front" and "back" are triply asymptotic triangles.

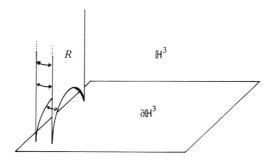

FIGURE 5.11.

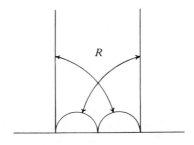

FIGURE 5.12.

When the line L, passing across the front triangle, crosses an edge to the back, it appears to have been "reflected" in the edge. Thus, L can be viewed as the trajectory of a billiard ball on a triply asymptotic triangular table.

Exercises

5.4.1. Show that the result of identifying edges of R as shown in Figure 5.12 is topologically a torus minus one point ("punctured torus").

5.4.2. Find explicit hyperbolic translations g^*, h^* which realize the identifications shown in Figure 5.12, and show that $\langle g^*, h^* \rangle$ is a free group.

5.4.3. Show that there are lines on the punctured torus with the property stated in the above theorem for lines on the punctured sphere.

5.5   General Construction of Hyperbolic Surfaces from Polygons

The constructions of Sections 5.2 to 5.4 are instances of a very general method for constructing hyperbolic surfaces. Any hyperbolic polygon can be converted to a hyperbolic surface by "pasting" edges together in pairs, provided obvious length and angle conditions are satisfied. To state these conditions precisely, we use the following definitions.

A *hyperbolic polygon* Π is a region of $\mathbb{H}^2$ bounded by a simple polygonal path of finitely many $\mathbb{H}^2$-line segments and segments of $\partial\mathbb{H}^2$, called *proper* and *improper* edges of Π, respectively. The endpoints of edges are called *vertices* of Π, with those in $\mathbb{H}^2$ being called *proper* (as opposed to *improper vertices*, which are the ends on $\partial\mathbb{H}^2$ of infinitely long edges). For example, we constructed the pseudosphere in Section 5.2 from the polygon Π whose edges are $\text{Re}(z) = -\pi$, $\text{Re}(z) = \pi$ and the portion of $\partial\mathbb{H}^2$ with $-\pi \leq$

$\text{Re}(z) \leq \pi$, and whose vertices are $-\pi$, π, ∞, all of which are improper. The proper edge $\text{Re}(z) = -\pi$ was "pasted" onto the proper edge $\text{Re}(z) = \pi$ by the hyperbolic isometry $t_{2\pi}$.

In general, an *edge pairing* of a polygon Π is a partition of the proper edges into pairs $\{e, e'\}$ of equal length (possibly infinite), with an $\mathbb{H}^2$-isometry $g_{e,e'} : e \to e'$ for each pair. Points $w \in e$ and $g_{e,e'}(w) = w' \in e'$ are said to be *identified* by the edge pairing. We also say that w is identified with w'' if w is identified with w', and w' with w''. Such a chain of identifications can occur with vertices, and we call a maximal set $\{v_1, \ldots, v_k\}$ of identified vertices a *vertex cycle*. An edge pairing of Π defines an *identification space* S_Π whose *points* are

(i) interior points u of Π,

(ii) pairs $\{w, w'\}$ of identified interior points of proper edges of Π, and

(iii) cycles $\{v_1, \ldots, v_k\}$ of proper vertices of Π. (Since an isometry sends points of $\mathbb{H}^2$ to points of $\mathbb{H}^2$, two identified vertices are either both proper or both improper.)

Theorem. *The identification space $S = S_\Pi$ has a distance function (agreeing with $\mathbb{H}^2$-distance for sufficiently small regions in the interior of Π) making it a hyperbolic surface when the angles of each vertex cycle sum to 2π.*

Proof. We define $d_S(A, B)$ as the inf of the lengths of all polygonal paths from A to B on S_Π. There is an obvious way to interpret a path p_{AB} from A to B on S_Π as a sequence of paths on Π, and hence assign it a length. Simply decompose p_{AB} into $p_{Aw_1}, p_{w_1'w_2}, p_{w_2'w_3}, \ldots, p_{w_nB}$ where

$w_1 = $ first point where p_{AB} crosses an edge of Π,

$w_1' = $ the point (identified with w_1) where p_{AB} reenters Π,

$$\vdots \qquad\qquad \vdots$$

$w_n = $ last point where p_{AB} reenters Π,

and let

$$\text{length}_S(p_{AB}) = \mathbb{H}^2\text{-length}(p_{Aw_1}) + \cdots + H^2\text{-length}(p_{w_nB}).$$

Then we define

$$d_S(A, B) = \inf\{\text{length}_S(p_{AB}) \mid \text{polygonal paths } p_{AB}\}.$$

The inf is obviously > 0 if A, B are distinct points of S_Π.

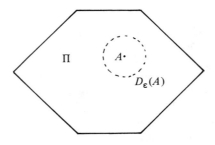

FIGURE 5.13.

We now show that for each $A \in S_\Pi$ there is an $\epsilon > 0$ such that

$$D_\epsilon(A) = \{B \in S \mid d_S(A, B) < \epsilon\}$$

is isometric to an $\mathbb{H}^2$-disc.

(i) If A is an interior point u of Π, choose ϵ less than half the minimum $\mathbb{H}^2$-distance from u to an edge of Π (Figure 5.13). Then $D_\epsilon(A)$ is precisely the open $\mathbb{H}^2$-disc $D_\epsilon(u)$ of radius ϵ and center u, and for $B, C \in D_\epsilon(A)$

$$d_S(B, C) = d_{\mathbb{H}^2}(B, C)$$

because the shortest path from B to C is necessarily the $\mathbb{H}^2$-line segment BC. (By choice of ϵ, any p_{BC} which goes through the boundary of Π is longer than the diameter of $D_\epsilon(u)$. Thus, $D_\epsilon(A)$ is isometric to the $\mathbb{H}^2$-disc $D_\epsilon(u)$ (under the identity map).

(ii) If A is a pair $\{w, w'\}$ of identified interior points of edges of Π, choose ϵ less than a quarter of the minimum distance from w or w' to the nearest vertex or other edge of Π. Then $D_\epsilon(A)$ is the union

$$D_\epsilon(\{w, w'\}) = (D_\epsilon(w) \cap \Pi) \cup (D_\epsilon(w') \cap \Pi)$$

of "half-discs" of radius ϵ and centers w, w' (Figure 5.14). (The broad arrow indicates the side-pairing isometry which sends w to w'; in this case, an orientation-reversing isometry.)

Then for $B, C \in D_\epsilon(A) = D_\epsilon(\{w, w'\})$, we have

$$d_S(B, C) = \begin{cases} D_{\mathbb{H}^2}(B, C) & \text{if } B, C \in \text{same half disc} \\ D_{\mathbb{H}^2}(B', C) & \text{if } B \in D_\epsilon(w), C \in D_\epsilon(w') \\ D_{\mathbb{H}^2}(B, C') & \text{if } B \in D_\epsilon(w'), C \in D_\epsilon(w) \end{cases}$$

because by choice of ϵ any p_{BC} which goes out of $D_\epsilon(\{w, w'\})$ and through the boundary of Π is longer than the diameter of $D_\epsilon(\{w, w'\})$. In other words, $D_\epsilon(A)$ is isometric to $D_\epsilon(w')$, under the map which is the identity on their common half and the side-pairing isometry on the other half [and $D_\epsilon(A)$ is similarly isometric to $D_\epsilon(w)$].

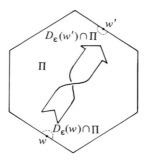

FIGURE 5.14.

(iii) If A is a vertex cycle $\{v_1, \ldots, v_k\}$, choose ϵ less than a quarter of the length of any edge of Π. Then $D_\epsilon(A)$ is the union

$$D_\epsilon(\{v_1, \ldots, v_k\}) = (D_\epsilon(v_1) \cap \Pi) \cup \ldots \cup (D_\epsilon(v_k) \cap \Pi))$$

of "sectors" $V_\epsilon(v_i) = D_\epsilon(v_i) \cap \Pi$ of radius ϵ and center v_i (Figure 5.15). When the angle sum of these sectors is 2π, each $D_\epsilon(v_j)$ can be decomposed into sectors isometric to the $V_\epsilon(v_i)$, adjacent to each other in the same cyclic order as that in which the $V_\epsilon(v_i)$ occur around the vertex $\{v_1, \ldots, v_k\}$ on S_Π. Moreover, for any $B, C \in D_\epsilon(A) = D_\epsilon(\{v_1, \ldots, v_k\})$

$$d_S(B, C) = d_{\mathbb{H}^2}(B^{(j)}, C^{(j)}),$$

where $B^{(j)}$, $C^{(j)}$ are the points in $D_\epsilon(v_j)$ corresponding to B, C under the correspondence of sectors just described. [This again follows by choice of ϵ, for reasons like those in (i) and (ii).] Thus, $D_\epsilon(A)$ is isometric to any $D_\epsilon(v_j)$. $\square$

Remark. The more elegant construction of the surface S_Π as an orbit space $\mathbb{H}^2/\Gamma$ is not open to us in this proof because we do not know, initially, that Γ exists. The Killing–Hopf theorem gives a Γ such that $S_\Pi = \mathbb{H}^2/\Gamma$ only when S_Π is *complete*, i.e., when each line on S_Π extends arbitrarily far. In Section 5.7 we shall show that this is always the case when the edges of Π are finite.

Exercise

5.5.1. By suitable choice of hyperbolic polygon, show that the sphere with $n \geq 4$ punctures is a hyperbolic surface.

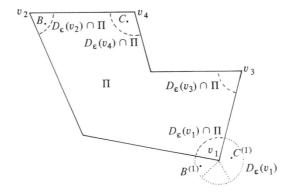

FIGURE 5.15.

5.6   Geometric Realization of Compact Surfaces

It is a theorem of topology (Radó [1924]) that any compact surface is homeomorphic to the identification space S_Π of a polygon Π whose edges are identified in pairs. Π is normally taken to be a euclidean polygon, but because the identifications are not required to be isometries—only homeomorphisms—Π could equally well be a spherical or hyperbolic polygon. The choice of geometry only becomes important when one asks whether each compact surface can be realized geometrically, i.e., as a spherical, euclidean, or hyperbolic surface. Assuming the theorem of Radó mentioned above, we shall show that the answer is yes.

Since the edges of Π are identified in pairs, the number of edges is even. If Π is a 2-gon, the two possible ways of identifying edges (Figure 5.16) give the sphere and the projective plane, respectively. (We are using the notation e, e' from Section 5.5 to indicate identified edges, with arrows to indicate the direction of identification.) Thus, any surface S_Π defined by a 2-gon Π can be realized as a spherical surface.

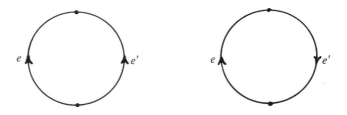

FIGURE 5.16.

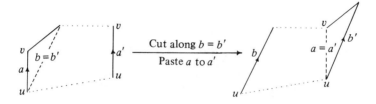

FIGURE 5.17.

If Π is a 4-gon, an enumeration of the possible identifications shows that S_Π is either a sphere, projective plane, torus or Klein bottle. Thus, S_Π can also be realized geometrically in this case because the torus and Klein bottle are euclidean surfaces. The interesting part is that all other surfaces S_Π can be realized as hyperbolic surfaces. To be able to construct the hyperbolic structure fairly directly, we use the following lemma, which enables us to assume that Π has a particularly simple edge pairing.

Lemma. *Any topological surface S_Π not homeomorphic to the sphere is homeomorphic to an S_{Π^*} for which Π^* has a single vertex cycle.*

Proof. Suppose that the polygon Π has at least two vertex cycles, $u = \{u_1, \dots, u_m\}$ and $v = \{v_1, \dots, v_n\}$. It will suffice to show how to convert Π to a polygon Π' for a homeomorphic surface with fewer vertices in v, and no new sets of identified vertices. To simplify notation we suppose all vertices in u to be labeled u, and all vertices in v to be labeled v.

Since not all vertices of Π are in u, there is an edge a of Π with one end u and the other end non-u, say v (without loss of generality). Let a' be the edge of Π identified with a. If a' is not the edge that meets a at v, then we can make the construction shown in Figure 5.17. The result is a polygon Π' with one less v vertex, and no new classes of equivalent vertices. Also, $S_{\Pi'}$ is homeomorphic to S_Π.

If a' is the edge which meets a at v, then a and a' must be oppositely directed (otherwise u, v become identified), and we can remove v as shown in Figure 5.18, unless a, a' are the only edges of Π, in which case S_Π is homeomorphic to a sphere. $\square$

Theorem. *Any compact surface can be realized geometrically.*

Proof. Having disposed of 2-gons and 4-gons in the remarks above, we can assume that the surface S_Π is defined by a polygon Π with $2n \geq 6$ edges. Also, by the lemma, we can assume that the edge pairing identifies all vertices of Π.

It then follows from Section 5.5 that S_Π can be given the structure of a hyperbolic surface if Π is chosen to be a hyperbolic polygon whose paired

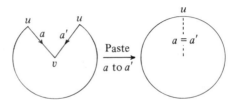

FIGURE 5.18.

FIGURE 5.19.

edges are of equal (hyperbolic) length and whose angle sum is 2π. The easiest way to do this is to let Π be a regular $2n$-gon with angle sum 2π.

We know from Exercise 4.7.2 that

$$\text{area}(2n\text{-gon}) = (2n - 2)\pi - \text{angle sum}.$$

Hence, we require

$$(2n - 2)\pi - \text{area}(\Pi) = 2\pi. \tag{1}$$

By taking the centre of Π to be 0 in $\mathbb{D}^2$, one sees that regular $2n$-gons Π exist with arbitrary hyperbolic diameter. It is also clear that area(Π) varies continuously with the diameter, between 0 (for diameter 0) and $(2n - 2)\pi$ (when the angles become zero). Since $2n > 4$, it follows from the intermediate value theorem that area(Π) takes a value which satisfies (1). With this area, the angle sum is 2π, as required.   □

Exercises

5.6.1. Show that the surface defined by the identifications of the hexagon shown in Figure 5.19 is homeomorphic to a torus. Is it a geometric torus?

5.6.2. Show that the surface shown in Figure 5.20 can be obtained identifying the edges of an octagon.

FIGURE 5.20.

5.7   Completeness of Compact Geometric Surfaces

Suppose that S_Π is a surface obtained by identifying sides of a polygon Π (spherical, euclidean, or hyperbolic). As mentioned in Section 5.5, the Killing–Hopf theorem enables us to express S_Π as a quotient ($\mathbb{S}^2/\Gamma$, $\mathbb{R}^2/\Gamma$, or $\mathbb{H}^2/\Gamma$) provided each line segment in S_Π can be extended indefinitely. This is not always the case for non-compact hyperbolic Π, as may be seen from Exercises 5.7.1 and 5.7.2. However, we can prove the following:

Theorem. *If Π is compact, then S_Π is complete.*

Proof. Each line segment in S_Π is the image of a sequence of line segments in Π under the identification map $\Pi \to S_\Pi$. Figure 5.21 shows an example of a line segment $L = L_1 \cup L_2 \cup L_3$ on a torus S_Π constructed from a rectangle Π. Continuation of a line segment on S_Π corresponds to a process on Π which extends a line segment until it hits an edge e of Π at point w, then resumes at the point w' of e' identified with w, in the direction that makes it straight in the disc neighborhood $D(\{w, w'\})$ of $\{w, w'\}$ on S_Π (and similarly if the line segment hits a vertex of Π).

It is clear that this process can be continued to form an infinite sequence L_1, L_2, $L_3, \ldots$ of line segments crossing Π from one edge to another. The question is whether the total length is infinite.

If not, then length(L_n) $\to 0$ as $n \to \infty$. This means that, for n sufficiently large, L_n has to remain within the neighborhood of a *single* vertex $\{v_1, \ldots, v_k\}$ of S_Π (because there is a positive lower bound to the lengths

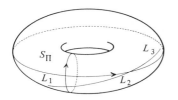

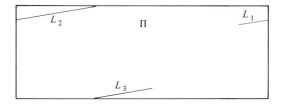

FIGURE 5.21.

of segments which pass from edge to edge outside fixed neighborhoods of vertices). But within a disc neighborhood of a vertex of S_Π, there can be at most finitely many consecutive segments of the same line—lying in the finitely many corners of Π which meet there—hence, we have a contradiction. $\qquad\qquad\qquad\qquad\qquad\qquad\qquad\qquad\qquad\qquad\qquad\qquad\quad$ $\square$

Exercises

5.7.1. Show that $\Pi = \{z \mid 1 \leq \operatorname{Re}(z) < 2\}$ is a fundamental region for the action of the group Γ generated by $g(z) = 2z$ on the *quarter* plane $\{z \mid \operatorname{Re}(z) > 0 \text{ and } \operatorname{Im}(z) > 0\}$.

5.7.2. Show that the surface S_Π constructed from the Π in Section 5.7.1 by identifying the sides via g is not complete.

5.8   Compact Hyperbolic Surfaces

The realization of identification spaces of polygons by geometric surfaces $\mathbb{S}^2/\Gamma$, $\mathbb{R}^2/\Gamma$, or $\mathbb{H}^2/\Gamma$ raises the following question: is it true, conversely, that a compact surface $\mathbb{S}^2/\Gamma$, $\mathbb{R}^2/\Gamma$, or $\mathbb{H}^2/\Gamma$ is the identification space of a polygon in the corresponding geometry? That is, does Γ have a polygon as a fundamental region? The answer is yes, except for the compact surface $\mathbb{S}^2$ itself. We already know how the projective plane, torus, and Klein bottle arise from polygons; hence, it remains to find polygonal fundamental regions for compact hyperbolic surfaces.

Theorem. *For any compact hyperbolic surface $\mathbb{H}^2/\Gamma$, there is a polygonal fundamental region for Γ.*

Proof. Since $\mathbb{H}^2/\Gamma$ is a surface, Γ is a group of hyperbolic isometries such that each $z \in \mathbb{H}^2$ has a neighborhood containing no other points in the Γ-orbit of z. For any $c \in \mathbb{H}^2$, the *Dirichlet region* $D(c)$ *with center* c consists of those points in $\mathbb{H}^2$ which are as near, or nearer, to c as they are to any other point in the Γ-orbit of c, i.e.,

$$D(c) = \{z \in \mathbb{H}^2 \mid d_{\mathbb{H}^2}(z,c) \leq d_{\mathbb{H}^2}(z,gc) \text{ for all } g \in \Gamma\},$$

where $d_{\mathbb{H}^2}$ denotes $\mathbb{H}^2$-distance. Each Γ-orbit obviously has at least one representative in $D(c)$, and in the interior of $D(c)$ (where the $<$ sign holds) at most one; hence, $D(c)$ is a fundamental region for Γ.

$D(c)$ is the intersection, over all $g \in \Gamma$, of the closed half-planes

$$H_g(c) = \{z \in \mathbb{H}^2 \mid d_{\mathbb{H}^2}(z,c) \leq d_{\mathbb{H}^2}(z,gc)\}.$$

Hence, $D(c)$ is a closed, convex region with boundary $\partial D(c)$ consisting of at most one segment from each of the lines

$$L_g(c) = \{z \in \mathbb{H}^2 \mid d_{\mathbb{H}^2}(z,c) = d_{\mathbb{H}^2}(z,gc)\}.$$

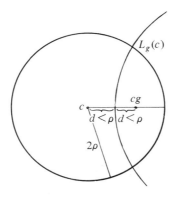

FIGURE 5.22.

(Recall from Section 4.3 that the set of points hyperbolically equidistant from two points is a hyperbolic line.) We have to show that only finitely many $L_g(c)$ actually met $\partial D(c)$.

This is where we use the assumption that $\mathbb{H}^2/\Gamma$ is compact. It implies that $D(c)$ is compact, and hence contained in a hyperbolic disc, of radius ρ, say. Now it follows from the definition of $L_g(c)$ that for each $L_g(c)$ which passes within distance ρ of c, there is a point gc in the Γ-orbit of c within distance 2ρ of c (Figure 5.22). Hence, if infinitely many $L_g(c)$ meet $D(c)$, the Γ-orbit of c contains infinitely many points within hyperbolic distance 2ρ of c. This implies that the Γ-orbit of c has an accumulation point, which contradicts the assumption that $\mathbb{H}^2/\Gamma$ is a surface.   □

Remark. To close the circle of ideas initiated by the construction of hyperbolic surfaces from polygons in Section 5.5, we should answer the question: what are the hyperbolic surfaces $\mathbb{H}^2/\Gamma$ obtained as identification spaces of arbitrary polygons (not necessarily compact)? The answer is that these are the surfaces for which Γ is *finitely generated*. It is clear from the construction of Γ as the covering isometry group—done in Section 2.9 for euclidean surfaces, but just the same for hyperbolic surfaces—that Γ is generated by the edge-pairing transformations we called $g_{e,e'}$ in Section 5.5. Since a polygon Π has only finitely many edges, by definition, this shows that Γ is finitely generated when Π is a fundamental region for Γ.

To prove the converse it suffices to construct a polygonal fundamental region for a given finitely generated Γ. This can be done along the lines of the theorem above, but is more complicated, so we refer to Beardon [1983, p. 255].

Exercise

5.8.1. Show that the region R in Section 5.3 is the Dirichlet region for the group Γ there.

5.9   Discussion

Like the euclidean plane, the hyperbolic plane admits "periodic" complex functions. The idea of periodicity can be generalized to hyperbolic geometry by calling a function f *periodic with respect to group* Γ if $f(g(z)) = f(z)$ for all $g \in \Gamma$. The most famous such group is the modular group discussed in Section 4.10. The associated periodic functions are called *elliptic modular functions* because they express the dependence of elliptic functions on the modulus ω_1/ω_2 (see Sections 2.10 and 4.10). Elliptic modular functions are not so easily constructed as the euclidean periodic functions such as e^z, but neither are they hopelessly difficult. Quite an elementary approach is given in Rademacher's little book [1983], and they are also accessible in the textbook of Jones and Singerman [1987] or the history of Gray [1986].

Elliptic modular functions have many marvelous properties—for example, Hermite [1858] used them to solve the general quintic equation—so they were studied long before their geometric meaning became apparent. A fundamental region for the modular group was first found by Dedekind [1877], yielding the famous tessellation shown in Figure 5.23. The union of any black and white region is a fundamental region for the modular group. (The division of the fundamental region into black and white halves allows us to show simultaneously the modular group, consisting of the isometries $z \mapsto \frac{\alpha z + \beta}{\gamma z + \delta}$, where α, β, γ, $\delta \in \mathbb{Z}$ and $\alpha\delta - \beta\delta = 1$, and the *extended* modular group in which $\alpha\delta - \beta\gamma = \pm 1$. Either half of a fundamental region for the modular group is a fundamental region for the extended modular group. In fact, the extended modular group is simply the modular group "extended by" a reflection which exchanges adjacent black and white regions. Conversely, the modular group is the "orientation-preserving subgroup" of the extended modular group. We shall generalize this black and white device to illustrate the relation between groups and their orientation-preserving subgroups in Chapter 7.)

Note that six such regions make up a fundamental region for the free group studied in Section 5.3, whose quotient surface is the triply punctured sphere. There is a complex function $\lambda(z)$, related to the elliptic modular functions, which is periodic with respect to this free group. It maps $\mathbb{H}^2$ onto the triply punctured sphere $\mathbb{C} - \{0, 1\}$ and, thus, realizes the universal covering of the triply punctured sphere (see Ford [1929, p. 157] and H. Cohn [1967, p. 312].

All the orientable quotients $\mathbb{H}^2/\Gamma$ admit nontrivial complex functions, analogous to the exponential and elliptic functions on the euclidean quo-

FIGURE 5.23.

tients $\mathbb{C}/\Gamma$. They are called *automorphic* functions, and their general theory was developed in the 1880s by Poincaré and Klein (see Poincaré [1985]). Just as elliptic functions stimulated the geometric and topological investigation of tori, automorphic functions stimulated the investigation of hyperbolic surfaces and led to such things as free groups and quasi-random geodesics. Another remarkable geometric discovery to emerge from complex function theory was the theorem of Schwarz [1879] that each compact hyperbolic surface has only finitely many isometries. (Contrast this with a torus $\mathbb{C}/\Gamma$, which has an isometry induced by each translation of $\mathbb{C}$.)

Quasi-random lines were first exhibited by Artin [1924] in yet another showcase for the modular group. The quotient of $\mathbb{H}^2$ by the modular group is not strictly a surface, as we have mentioned before (Section 4.10), but it does possess lines. They can be viewed as trajectories on a hyperbolic billiard table in the shape of a fundamental region in Figure 5.23. Our account of lines on the punctured sphere adapts Artin's construction to a true surface, using also some ideas of Nielsen [1925], who showed that there are quasi-random lines on all compact hyperbolic surfaces. For an engaging account of Artin's original construction, see Series [1982].

6
Paths and Geodesics

6.1 Topological Classification of Surfaces

With the Killing–Hopf theorem, that every geometric surface is of the form $\mathbb{S}^2/\Gamma$, $\mathbb{R}^2/\Gamma$, or $\mathbb{H}^2/\Gamma$, the problem of classifying surfaces is replaced by the problem of classifying groups Γ. In the spherical and euclidean cases this problem is easy to solve, as we have seen in Chapters 2 and 3, because there are only a small number of possibilities. However, in the hyperbolic case the number of possibilities is infinite, and the problem is best clarified by taking a new viewpoint, halfway between geometry and group theory. This is the viewpoint of topology briefly mentioned in Section 5.6.

In Section 5.6 we saw that each compact topological surface may be obtained as the identification space S_Π of a polygon Π. Moreover, if S_Π is not homeomorphic to $\mathbb{S}^2$, then Π may be chosen to have a single vertex cycle. We now refine this idea so as to classify the nonspherical surfaces S_Π by showing that all such surfaces are obtainable from *normal form* polygons Π. The normal form S_Π, as we shall eventually show in Section 6.7, are topologically distinct surfaces, and hence certainly geometrically distinct.

Actually, we shall carry out this program only for *orientable* surfaces (those for which any two identified edges in the boundary of Π are oppositely directed). The case of the nonorientable surfaces, which is similar but a little messier, is left to the exercises. The result is that each compact orientable surface is homeomorphic to one of the surfaces in Figure 6.1. These surfaces are distinguished informally by the number of "holes" or "handles", which is formalized by the concept of *genus*. The definition of genus is part of the following description of the normal forms.

6.1.1 Normal Form for Compact Orientable Surfaces

From now on we denote identified edges a, a' in the boundary of Π by the *same* letter, a, but when the edges are oppositely directed, we denote them by a, a^{-1}. The boundary path of Π is then a sequence of letters in which each letter appears twice, possibly with opposite exponents.

Normal Forms. Each compact orientable surface is homeomorphic to an identification space S_Π, where Π is a polygon with boundary path of the

FIGURE 6.1.

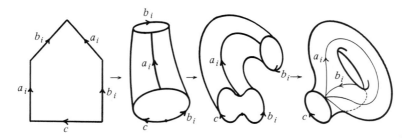

FIGURE 6.2.

form aa^{-1} or $a_1 b_1 a_1^{-1} b_1^{-1} \ldots a_n b_n a_n^{-1} b_n^{-1}$. In the former case, S_Π is said to be of *genus* 0, in the latter case *genus* n.

Remarks. (1) The genus n can be identified with the number of handles because each segment $a_i b_i a_i^{-1} b_i^{-1}$ in the boundary of Π gives rise to a handle, as Figure 6.2 shows.

(2) As already mentioned, our eventual aim is to show that different normal form surfaces are nonhomeomorphic, hence geometrically distinct. It is also plausible, though laborious to prove, that surfaces with the same normal form *are* homeomorphic. The main problem is to prove that any two polygons are homeomorphic. Details of this may be found in Moise [1977, p. 26].

Construction of Normal Forms. We can assume, by Section 5.6, that if the boundary of Π is not of the form aa^{-1}, then Π has just one equivalence class of identified vertices. Since S_Π is orientable, any two identified edges are oppositely directed in the boundary of Π. It follows that if a, a^{-1} is one pair of identified edges, then the remaining portions p_1, p_2 of the boundary contain an identified pair b, b^{-1} as shown in Figure 6.3. If, on the contrary, no edge in p_1 was identified with an edge in p_2, then the terminus of a could not be identified with its origin, contrary to the hypothesis that all vertices of Π are identified.

From the "separating pairs" of edges a, a^{-1} and b, b^{-1} we now create a handle $a_i b_i a_i^{-1} b_i^{-1}$ as follows:

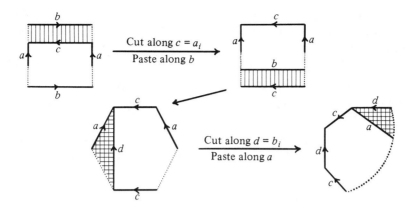

Note that any handles $a_j b_j a_j^{-1} b_j^{-1}$ already present in the boundary are not disrupted by this process. Hence, we can repeat it until we have a polygon Π' whose boundary consists entirely of handles. $S_{\Pi'} = S_{\Pi}$ because the "cutting" and "pasting" are nothing more than drawing new lines on S_{Π} and erasing old ones.

Exercises

6.1.1. Let $S_{\Pi_1,\ldots,\Pi_k}$ be a surface obtained by identifying edges of polygons $\Pi_1, \ldots, \Pi_k$ in pairs. Call $S_{\Pi_1,\ldots,\Pi_k}$ *orientable* if the boundaries of $\Pi_1, \ldots, \Pi_k$ can be oriented in such a way that each edge on $S_{\Pi_1,\ldots,\Pi_k}$ appears with both orientations among the face boundaries. Show that orientability is preserved by "cutting" and "pasting".

6.1.2. Let S_{Π} be a nonorientable surface (so there is at least one pair of identified edges with the same orientation in the boundary of Π) with a single vertex. Show that any pair of identified like-oriented edges in the boundary of Π can be replaced by a pair $c_i c_i$ of *adjacent* like-oriented edges in the boundary of Π. (Hint: One cut and one paste suffices.)

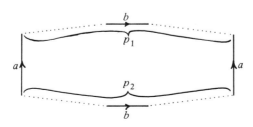

FIGURE 6.3.

6.1.3. Deduce from Exercise 6.1.2 and the handle contruction above that a nonorientable surface takes the form S_Π where the boundary of Π consists of handles and terms $c_i c_i$.

6.1.4. Show that a portion of the boundary of Π of the form $ccaba^{-1}b^{-1}$ may be replaced by $d^{-1}b^{-1}a^{-1}d^{-1}a^{-1}b^{-1}$ (one cut and one paste) and that the latter may be replaced by $c_1 c_1 c_2 c_2 c_3 c_3$ by applying Exercise 6.1.2 three times in a suitable order.

6.1.5. Conclude that each compact nonorientable surface has a normal form S_Π, where Π is a polygon with boundary of the form $c_1^2 c_2^2 \dots c_n^2$.

6.2   Geometric Classification of Surfaces

When the normal form orientable surfaces of Section 6.1 are realized geometrically, by the construction of Section 5.6, then any surface of genus 0 becomes a sphere, any surface of genus 1 becomes a torus, and surfaces of genus > 1 become hyperbolic. Conversely, we know from the classification of spherical and euclidean surfaces that the sphere is the *only* orientable spherical surface, and that all compact orientable euclidean surfaces are tori, hence of genus 1. To complete this train of thought, we would like to show that all compact orientable hyperbolic surfaces are of genus > 1. It would be even nicer to find some geometric property to distinguish the various genera $n > 1$.

Suppose then that $S = \mathbb{H}^2/\Gamma$ is a compact hyperbolic surface. By Section 5.7 we can realize S as the identification space S_Π of a convex hyperbolic polygon Π (the Dirichlet region for the group Γ). If Π is an m-gon, then we know from Section 4.7 that

$$\text{area}(S_\Pi) = \text{area}(\Pi) = (m-2)\pi - \text{angle sum of } \Pi$$
$$= (m-2)\pi - 2\pi(\text{number of vertices of } S_\Pi)$$

because the angles of Π are grouped into sums of 2π around the vertices of S_Π. We now let

$$V(S_\Pi) = \text{number of vertices of } S_\Pi,$$

$$E(S_\Pi) = \text{number of edges of } S_\Pi, \text{ and}$$

$$F(S_\Pi) = \text{number of faces of } S_\Pi = 1.$$

Then we can rewrite $\text{area}(S_\Pi)$ as follows:

$$\begin{aligned}
\text{area}(S_\Pi) &= (2E(S_\Pi) - 2)\pi - 2\pi V(S_\Pi) \\
&= 2\pi(-V(S_\Pi) + E(S_\Pi) - F(S_\Pi)) \quad \text{since } F(S_\Pi) = 1 \qquad (1) \\
&= -2\pi\chi(S_\Pi),
\end{aligned}$$

where $\chi(S_\Pi) = V(S_\Pi) - E(S_\Pi) + F(S_\Pi)$ is called the *Euler characteristic* of S_Π. Equation (1) shows that the Euler characteristic of a compact hyperbolic surface is proportional to area, and that the constant of proportionality is -2π. It follows that these surfaces have *negative* Euler characteristic.

We now observe that Euler characteristic is invariant under the cutting and pasting operations used to reduce a surface to normal form in Sections 5.6 and 6.1. Cutting simultaneously increases E and F by 1; pasting simultaneously decreases them by 1 (or in the exceptional cases where adjacent edges a, a^{-1} are pasted together, it decreases V and E by 1). But the normal form

$$aa^{-1} \text{ has } \chi = 2 \ (V = 2, \ E = 1, \ F = 1), \text{ and}$$

$$a_1 b_1 a_1^{-1} b_1^{-1} \ldots a_n b_n a_n^{-1} b_n^{-1} \text{ has } \chi = 2 - 2n \ (V = 1, \ E = 2n, \ F = 1),$$

hence, the only compact orientable surfaces with negative χ are those of genus $n > 1$, as required.

The invariance of Euler characteristic is remarkably useful. Not only does it enable us to determine the genus n of a surface without reducing to normal form, by the formula

$$\chi = 2 - 2n \tag{2}$$

(which is also true for genus $n = 0$), it also relates genus to area for hyperbolic surfaces via (1) and (2). Thus, area is the geometric property that distinguishes the various genera of compact orientable hyperbolic surfaces.

We summarize our conclusions in the following theorem, which shows that the topological classification of surfaces can be restated as a geometric classification.

Theorem. *If S is a compact orientable surface of constant curvature $\kappa_S = 1$, 0 or -1, then*

  (i) *S has genus $0 \Rightarrow \kappa_S = 1$,*

  (ii) *S has genus $1 \Rightarrow \kappa_S = 0$, and*

  (iii) *S has genus $n > 1 \Rightarrow \kappa_S = -1$ and* area$(S) = 4\pi(n-1)$.

Exercises

6.2.1. Show that the projective plane has Euler characteristic 1.

6.2.2. Show that the polygon with boundary $a_1 a_2 \ldots a_n a_1^{-1} a_2^{-1} \ldots a_n^{-1}$ gives an arbitrary orientable surface for suitable choice of n, and similarly that $a_1^{-1} a_2 \ldots a_n a_1^{-1} a_2^{-1} \ldots a_n^{-1}$ gives an arbitrary nonorientable surface.

6.2.3. Interpret Figure 6.4 as a covering map. Hence, conclude that any nonorientable surface is (doubly) covered by an orientable surface. What

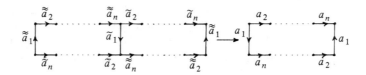

FIGURE 6.4.

does this imply for the area and Euler characteristic of the nonorientable surface?

6.2.4. Illustrate Exercises 6.2.2 and 6.2.3 with the example of the torus and the Klein bottle.

6.3   Paths and Homotopy

The invariance of Euler characteristic, and hence genus, under cutting and pasting suggests that it is a *topological* invariant. To prove this, we need to show that it can be defined in topological terms, that is, in terms of continuous functions. The key concept is that of a "path", which can be viewed intuitively as the continuous motion of a point.

A *path* p in a surface S is a continuous function $p : [0, 1] \to S$, where $[0, 1]$ is the unit interval of $\mathbb{R}$. The *endpoints* $p(0)$ and $p(1)$ of p are called, respectively, its *origin* and *terminus*.

It is helpful to interpret $[0, 1]$ as a time interval, so $p(x)$ is the position of the moving point at time x.

The topology of a surface is reflected by the variety of paths within it. In the plane, for example, any two paths p, q between points A and B are "topologically the same" in the sense that p is "deformable" into q with origin and terminus fixed. On the cylinder, however, the paths p_1, p_2 shown in Figure 6.5 are topologically different, and the "difference" is in some sense a path which winds once around the cylinder. To prepare for rigorous proofs of these results, we formalize the notion of a "deformation" from p to q as follows:

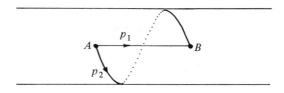

FIGURE 6.5.

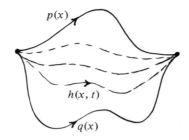

FIGURE 6.6.

Paths $p, q : [0, 1] \rightarrow S$ with the same origin and terminus are called
homotopic (in S) if there is a continuous function $h : [0, 1] \times [0, 1] \rightarrow S$,
called a *homotopy between p and q*, such that

(i) $h(x, 0) = p(x)$ and $h(x, 1) = q(x)$, and

(ii) $h(0, t) = p(0) = q(0)$ and $h(1, t) = p(1) = q(1)$ for all t.

Now it is helpful to think of x as "length" (in the preimage unit interval)
and t as "time". Then the function $h(x, t)$ is the path into which p has
been "deformed" at time t (Figure 6.6). Condition (i) says that the path
is p at time $t = 0$ and q at time $t = 1$, whereas condition (ii) says that the
endpoints are fixed throughout.

It follows immediately from the definition that the relation "p is homotopic to q" is an equivalence relation: p is homotopic to p via $h(x, t) = p(x)$;
if p is homotopic to q via $h(x, t)$, then q is homotopic to p via $h(x, 1 - t)$;
if p is homotopic to q via $f(x, t)$ and q is homotopic to r via $g(x, t)$, then
p is homotopic to r via

$$h(x, t) = \begin{cases} f(x, 2t) & \text{for } 0 \le t \le \frac{1}{2}, \\ g(x, 2t - 1) & \text{for } \frac{1}{2} \le t \le 1. \end{cases}$$

In view of this, to prove that any two paths in $\mathbb{R}^2$ with the same endpoints
are homotopic, it suffices to prove the following:

Proposition. *A path p in $\mathbb{R}^2$, with origin A and terminus B, is homotopic
to the line segment AB.*

Proof. If $A \ne B$, then by suitable choice of coordinate axes and scale
we can take $A = (0, 0)$ and $B = (1, 0)$. For each $(x, 0) \in AB$ consider
the line segment L_x from $(x, 0)$ to $p(x) = (p_1(x), p_2(x))$ (Figure 6.7). Our
"deformation" of p to AB slides each point $p(x)$ along L_x at constant speed
to $(x, 0)$ in unit time. That is, $h(x, t) = (p_1(x) + t(x - p_1(x)), p_2(x) - tp_2(x))$.

If $A = B$, we use the same construction, but with $A = B = (0, 0)$. In
this case, p is homotopic to the "constant path" $p_A(x) = A$. $\qquad\square$

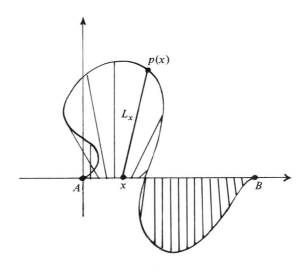

FIGURE 6.7.

The set of paths homotopic to p, in a given surface S, is called the *homotopy class* of p and denoted by $[p]$.

Thus, the proof of the proposition shows that each homotopy class $[p]$ in $\mathbb{R}^2$ contains a line segment. Since the line segment AB between points $A, B \in \mathbb{R}^2$ is unique, there is a unique line segment in $[p]$, called the *geodesic representative* of $[p]$. The existence and uniqueness of geodesic representatives, for suitably generalized homotopy classes, will be proved for other surfaces later in this chapter. It leads to a classification of closed straight lines on compact surfaces (Sections 6.8 and 6.9). The first step is to settle the homotopy properties of $\mathbb{S}^2$, $\mathbb{R}^2$ and $\mathbb{H}^2$.

Theorem. *In $\mathbb{S}^2$, $\mathbb{R}^2$, and $\mathbb{H}^2$, any paths with the same endpoints are homotopic.*

Proof. Since the homotopy relation is an equivalence, it will suffice to show that any path is homotopic to a line segment between its endpoints (unique for $\mathbb{R}^2$, $\mathbb{H}^2$, some fixed choice for $\mathbb{S}^2$). We have already done this for $\mathbb{R}^2$ in the proposition above. It follows immediately for $\mathbb{H}^2$ by using the $\mathbb{D}^2$-model and viewing $\mathbb{D}^2$ as part of $\mathbb{R}^2$ with the real axis of $\mathbb{D}^2$ as part of the x-axis of $\mathbb{R}^2$.

It also follows for $\mathbb{S}^2$ in the special case where the path p does not map onto all of $\mathbb{S}^2$ because in this case one can stereographically project $\mathbb{S}^2$ onto $\mathbb{R}^2$ from a point not on p, and again reduce to the proposition for $\mathbb{R}^2$.

However, space-filling curves exist, and if p is onto $\mathbb{S}^2$, we first decompose p into finitely many subpaths which are *not* onto $\mathbb{S}^2$. This is possible because $p : [0,1] \to \mathbb{S}^2$ is continuous, hence uniformly continuous (since $[0,1]$ is closed). In particular, we can decompose $[0, 1]$ into n subintervals $[0, \frac{1}{n}]$, $[\frac{1}{n}, \frac{2}{n}], \ldots, [\frac{n-1}{n}, 1]$, each of which is mapped by p into, say, a hemisphere. We can then use the argument of the special case to deform each subpath $p : [\frac{i}{n}, \frac{i+1}{n}] \to \mathbb{S}^2$ into a line segment, so that $p : [0, 1] \to \mathbb{S}^2$ is homotopic to a (spherical) polygon q. Since a polygon does not fill $\mathbb{S}^2$, we can use the special case again to show that q, and hence p, is homotopic to a line segment. $\qquad\square$

Remark. The proof for the $\mathbb{S}^2$ case is the first place where *compactness* of paths becomes important. Similar applications of compactness—to argue that a path or homotopy can be decomposed into a finite number of "small" pieces—will be important in the next section.

Exercise

6.3.1. If p is a path from A to B, give a suitable definition of the inverse path p^{-1} from B to A, and hence define a homotopy between pp^{-1} and the constant path $p(x) = A$.

6.4   Lifting Paths and Lifting Homotopies

The difference between the (presumed) nonhomotopic paths p_1, p_2 on the cylinder C, shown in Figure 6.5, is seen in a new light when we view p_1, p_2 as the "projections" $\Gamma\tilde{p}_1$, $\Gamma\tilde{p}_2$ of paths $\tilde{p}_1$, $\tilde{p}_2$ on $\mathbb{R}^2$ under the orbit map $\Gamma\cdot : \mathbb{R}^2 \to \mathbb{R}^2/\Gamma = C$ (Figure 6.8). If $\tilde{p}_1$, $\tilde{p}_2$ have the same origin $\tilde{A}$, they have *different* termini $\tilde{B}^{(1)}$, $\tilde{B}^{(2)}$. In fact, $\tilde{B}^{(2)} = g(\tilde{B}^{(1)})$ for some nonidentity $g \in \Gamma$ because $\Gamma\tilde{B}^{(1)} = \Gamma\tilde{B}^{(2)}$, and the element g "measures" in some sense the "topological difference" between p_1 and p_2.

In general, we measure the topological difference between paths p_1, p_2 with the same endpoints O, P on a surface $S = \tilde{S}/\Gamma$, where $\tilde{S}$ is $\mathbb{S}^2$, $\mathbb{R}^2$, or $\mathbb{H}^2$, by "lifting" p_1, p_2 to paths $\tilde{p}_1$, $\tilde{p}_2$ with the same origin $\tilde{O}$ on $\tilde{S}$, and comparing the termini $\tilde{P}^{(1)}$ and $\tilde{P}^{(2)} = g(\tilde{P}^{(1)})$. The element $g \in \Gamma$ represents the topological difference between p_1 and p_2. This idea works thanks to the following *path lifting* and *homotopy lifting* properties.

Path Lifting Property. *For each path p in S with origin O, there is a unique path $\tilde{p}$ in $\tilde{S}$ with given origin $\tilde{O}$ (lying over O) such that $p = \Gamma\tilde{p}$. We call $\tilde{p}$ the lift of p with origin $\tilde{O}$.*

Proof. It will suffice to partition $[0, 1]$ into finitely many subintervals $[x_i, x_{i+1}]$ small enough that p maps $[x_i, x_{i+1}]$ into an open disc $D_i \subset S$, i.e., the bijective image under $\Gamma\cdot$ of a disc $\tilde{D}_i \subset \tilde{S}$, because it then follows by induction on i that $p : [0, x_i] \to S$ lifts uniquely to $\tilde{p} : [0, x_i] \to \tilde{S}$ with $\tilde{p}(\tilde{O}) = O$ and $\Gamma\tilde{p} = p$. (The portion p_1 of p in D_1 lifts uniquely to a $\tilde{p}_1$ in the unique disc $\tilde{D}_1 \subset \tilde{S}$ such that $\Gamma\tilde{D}_1 = D_1$ and $\tilde{O} \in \tilde{D}_1$ by inverting the

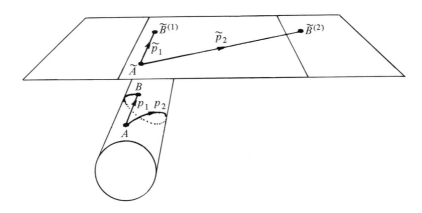

FIGURE 6.8.

bijection $\Gamma \cdot : \tilde{D}_1 \to D_1$. Then the portion p_2 of p in D_2 lifts uniquely to a $\tilde{p}_2$ in the unique disc $\tilde{D}_2 \subset \tilde{S}$ that contains the terminus of $\tilde{p}_1$ and such that $\Gamma \tilde{D}_2 = D_2$, and so on.)

The finitely many intervals $[x_i, x_{i+1}]$ are constructed by repeated bisection. If p does not map $[0,1]$ into an open disc $D \subset S$, we bisect $I = [0,1]$ into $I_0 = [0, \frac{1}{2}]$ and $I_1 = [\frac{1}{2}, 1]$, and ask whether p maps I_0 or I_1 into an open disc. Whenever this fails to happen for I_i, we bisect I_i into I_{i0} and I_{i1} and ask the same question.

This process eventually terminates. If not, we get a nested sequence of intervals $I \supset I' \supset I'' \supset \ldots$, none of which is mapped into an open disc by p, and with a single common point x. But by continuity, p maps *some* open interval containing x into an open disc $D \subset S$, so we have a contradiction. Since the process always terminates, we get $[0,1]$ divided into finitely many subintervals $[x_i, x_{i+1}]$ with the required property.   □

The path lifting property enables us to associate a unique point $\tilde{P} \in \tilde{S}$ with each path p in S with origin O, namely, the terminus of the unique lift $\tilde{p}$ of p with origin $\tilde{O} \in \tilde{S}$. We now show that $\tilde{P}$ depends only on the homotopy class of p.

Homotopy Lifting Property. *Paths p_1, p_2 with origin O in S are homotopic if and only if their lifts $\tilde{p}_1$, $\tilde{p}_2$ with origin $\tilde{O} \in \tilde{S}$ are homotopic.*

Proof. A homotopy $\tilde{h}$ between $\tilde{p}_1$ and $\tilde{p}_2$ obviously yields a homotopy, $\Gamma \cdot \tilde{h}$, between p_1 and p_2. Conversely, suppose $h : [0,1] \times [0,1] \to S$ is a homotopy between p_1 and p_2. By repeated bisections, the square $[0,1] \times [0,1]$ can be divided into a finite number of subsquares, each of which is mapped by h into a disc of S. We can then lift h to a homotopy $\tilde{h} : [0,1] \times [0,1] \to \tilde{S}$ between $\tilde{p}_1$ and $\tilde{p}_2$ by a process like that used to lift p to $\tilde{p}$ earlier.   □

Unique path lifting and homotopy lifting together give the following:

Theorem. *If p is a path in S with origin O, and if $\tilde{p}$ denotes the unique lift of p with origin $\tilde{O} \in \tilde{S}$, then*

$$[p] \mapsto terminus \ of \ \tilde{p}$$

defines a bijection between the homotopy classes $[p]$ and the points of $\tilde{S}$, where $\tilde{S} = \mathbb{S}^2$, $\mathbb{R}^2$ or $\mathbb{H}^2$.

Proof. By the homotopy lifting property, each p' homotopic to p lifts to a $\tilde{p}'$ homotopic to $\tilde{p}$, hence with the same terminus as $\tilde{p}$, so the map

$$[p] \mapsto \text{terminus of } \tilde{p}$$

is well-defined.

The map is onto $\tilde{S}$ because for any point $\tilde{P} \in \tilde{S}$ we can take a line segment $\tilde{O}\tilde{P}$ from $\tilde{O}$ to $\tilde{P}$ and let $p = \Gamma(\tilde{O}\tilde{P})$.

To show that the map is one-to-one, suppose that p_1, p_2 are two paths in S with origin O and that their lifts $\tilde{p}_1$, $\tilde{p}_2$ with origin $\tilde{O}$ have the same terminus. Then $\tilde{p}_1$ is homotopic to $\tilde{p}_2$ by the theorem in Section 6.3, and hence p_1 is homotopic to p_2 (by the easy direction of the homotopy lifting property), as required. $\square$

Remark. This theorem shows that the universal cover $\tilde{S}$ of S has a topological meaning—it is the space of homotopy classes $[p]$ of paths p with a fixed origin O on S. These classes can be described more picturesquely as "points with tails" (I owe this description to a manuscript by Michio Kuga entitled *Galois' Dream*). The homotopy class $[p]$ corresponds to a point $P \in S$ (the terminus of p) with an elastic "tail" tied to O (a homotopy class of paths from O to P).

Exercise

6.4.1. Show (without assuming $\tilde{S}$ is $\mathbb{S}^2$, $\mathbb{R}^2$, or $\mathbb{H}^2$) that any two paths with the same endpoints on the universal cover are homotopic.

6.5 The Fundamental Group

Among the homotopy classes $[p]$ of paths on S with origin O, those of the *closed* paths are of particular interest. One such class is the *trivial class* of paths p homotopic to the point O [given formally by the constant path $p(x) = O$]. The *nontrivial* homotopy classes, if any, reflect the topological difference between S and its universal cover $\tilde{S}$, on which only the trivial homotopy class exists.

The topological structure of S can be understood in more detail by constructing a group from the homotopy classes of closed paths with origin O, called the *fundamental group* $\pi_1(S)$ of S. The *product of classes* $[p_1]$, $[p_2]$ is defined by

$$[p_1][p_2] = [p_1 p_2],$$

where $p_1 p_2$ denotes the path p_1 "followed by" the path p_2. The path $p_1 p_2$ can be defined formally by

$$(p_1 p_2)(x) = \begin{cases} p_1(2x) & \text{for } 0 \le x \le \frac{1}{2}, \\ p_2(2x - 1) & \text{for } \frac{1}{2} \le x \le 1. \end{cases}$$

It is easy to verify directly that the classes $[p]$ form a group under this operation—for example, Exercise 6.3.1 gives the existence of inverses—however, we will obtain this fact as a consequence of the following more interesting result.

Theorem. *If $S = \tilde{S}/\Gamma$, where $\tilde{S}$ is $\mathbb{S}^2$, $\mathbb{R}^2$, or $\mathbb{H}^2$, then $\pi_1(S) \cong \Gamma$.*

Proof. Since the map $[p] \mapsto$ terminus of $\tilde{p}$ is a bijection between the homotopy classes of paths p in S with origin O and points $\tilde{P}$ on $\tilde{S}$ (by Section 6.4), its restriction to $[p]$ for closed p gives a bijection between $\pi_1(S)$ and the set of $\tilde{P} \in \tilde{S}$ such that $\Gamma\tilde{P} = O$. The latter set is simply the Γ-orbit of $\tilde{O}$, by definition of $S = \tilde{S}/\Gamma$, and hence its points can be associated with elements $g \in \Gamma$. The natural way to do this is to associate $g\tilde{O}$ with g.

Thus, we have a bijection

$$\pi_1(S) \ni [p] \leftrightarrow g \in \Gamma, \quad \text{where } g\tilde{O} = \text{terminus of } \tilde{p}.$$

It remains to show that products of homotopy classes correspond to products of elements of Γ under this bijection. That is, if $\tilde{p}_1$ terminates at $g_1\tilde{O}$ and $\tilde{p}_2$ terminates at $g_2\tilde{O}$, then $\widetilde{p_1 p_2}$ terminates at $g_1 g_2\tilde{O}$. Since $\widetilde{p_1 p_2}$ consists of the lift $\tilde{p}_1$ of p_1 (with origin O) followed by the lift $\tilde{p}_2'$ of p_2 with origin $\tilde{O}' = $ terminus $g_1\tilde{O}$ of $\tilde{p}_1$, it will suffice to show that the lift $\tilde{p}_2'$ of p_2 with origin $g_1\tilde{O}$ terminates at $g_1 g_2\tilde{O}$.

By definition of g_2, the lift $\tilde{p}_2$ of p_2 with origin $\tilde{O}$ ends at $g_2\tilde{O}$. The lift $\tilde{p}_2'$ is simply $g_1\tilde{p}_2$ because the latter has origin g_1 (origin $\tilde{p}_2$) $= g_1\tilde{O}$ and $\Gamma g_1\tilde{p}_2 = \Gamma\tilde{p}_2 = p_2$; hence its terminus is g_1 (terminus $\tilde{p}_2$) $= g_1 g_2\tilde{O}$ as required. $\square$

Corollary. *If S is a complete geometric surface with $\pi_1(S) = \{1\}$, then S is $\mathbb{S}^2$, $\mathbb{R}^2$, or $\mathbb{H}^2$.*

Proof. By the Killing–Hopf theorem, $S = \tilde{S}/\Gamma$ where $\tilde{S}$ is $\mathbb{S}^2$, $\mathbb{R}^2$, or $\mathbb{H}^2$. By the theorem above, $\Gamma = \pi_1(S)$; hence $S = \tilde{S}/\{1\} = \tilde{S}$. $\square$

Remarks. (1) Note how this proof gives us another justification, as promised in Section 1.2, for taking the product $g_1 g_2$ of isometries to be a substitution of g_2 in g_1. It is substitution that corresponds to the natural product of paths. The isomorphism between $\pi_1(S)$ and Γ means that, when convenient, we can interpret an isometry $g \in \Gamma$ as a homotopy class on S, or even as a particular curve in the class. To minimize confusion we shall speak of the "isometry g", "curve g", etc., to indicate which interpretation we have in mind.

(2) If, following the remark in Section 6.4, we view $\tilde{S}$ as a space constructed from S, we can lift neighborhoods from S to $\tilde{S}$ and hence show $S = \tilde{S}/\pi_1(S)$. Thus, even when a surface S is not given a priori as a quotient, we can prove directly that it is the quotient of its universal cover by its fundamental group.

(3) Since $\pi_1(S)$ is defined in terms of certain continuous maps into S (paths and homotopies), $\pi_1(S) \cong \pi_1(S')$ for any S' homeomorphic to S. That is, $\pi_1(S)$ is a topological invariant of S. This means in particular that if S, S' are surfaces with $\pi_1(S) \not\cong \pi_1(S')$, then S, S' are not homeomorphic.

We can use Remark (3) to prove the topological distinctness of closed orientable surfaces of different genus, claimed in Section 6.1. Certainly, because $\pi_1(\mathbb{S}^2) = \{1\}$ by Section 6.3 and $\pi_1(\text{torus}) = \mathbb{Z}^2$ by Section 2.3, we have a proof that $\mathbb{S}^2$ and the torus are not homeomorphic. We can also see that neither $\mathbb{S}^2$ nor the torus is homeomorphic to any compact surface $S = \mathbb{H}^2/\Gamma$ of higher genus, e.g., by showing that $\pi_1(S) = \Gamma$ is not abelian. However, to distinguish the various surfaces of genus > 1 from each other, we need more detailed information about their fundamental groups. This information will be obtained in the next section.

Exercises

6.5.1. Show that $\pi_1(\text{projective plane}) = \mathbb{Z}_2$, whereas π_1 (orientable surface of genus ≥ 1) has no elements of finite order.

6.5.2. Show that $\pi_1(S)$ is independent (up to isomorphism) of the choice of origin O on S.

6.6 Generators and Relations for the Fundamental Group

Recall from Section 6.1 that an orientable surface of genus $n \geq 1$ can be given the normal form S_Π where Π is a polygon with boundary path $a_1 b_1 a_1^{-1} b_1^{-1} \ldots a_n b_n a_n^{-1} b_n^{-1}$. Also, by Section 5.6, S_Π becomes a geometric surface when Π is constructed in the appropriate geometry. When $n = 1$, Π can be taken to be a square in $\tilde{S}_\Pi = \mathbb{R}^2$, the edge identifications can be realized by translations (Figure 6.9) and S_Π is a euclidean surface. When $n > 1$, Π can be taken to be a $4n$-gon in $\tilde{S}_\Pi = \mathbb{H}^2$ with angle sum 2π and the edge identifications can be realized by hyperbolic isometries (which are in fact translations). Figure 6.10 shows the octagon for the $n = 2$ case,

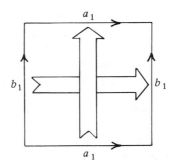

FIGURE 6.9.

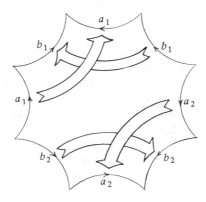

FIGURE 6.10.

in the $\mathbb{D}^2$-model. Each edge of the octagon maps to a closed curve on S_Π, namely, the join of like-labeled edges on Π. We shall use a_i (resp. b_i) as a label for all lifts $\tilde{a}_i$, $\tilde{a}'_i$, $\tilde{a}''_i$, ... (resp. $\tilde{b}_i$, $\tilde{b}'_i$, $\tilde{b}''_i$, ...) of this curve on $\tilde{S}_\Pi$, as a name for the curve itself on S_Π, *and* as an abbreviation for the homotopy class $[a_i]$ (resp. $[b_i]$). (The latter follows the usual mathematical practice of confusing an equivalence class with an "obvious" representative.) It may help to recall, from Section 6.5, that a_i (resp. b_i) can also be viewed as the isometry in Γ that maps the initial point of each curve labeled a_i (resp. b_i) to its terminal point.

Let $\tilde{G}$ be the labeled graph on $\tilde{S}_\Pi$ consisting of all lifts of the curves a_i, b_i, with their labels, and arrows indicating direction. We already know, from Section 6.5, that the vertices of this diagram are the points of the form $g\tilde{O}$, where $\tilde{O} \in \tilde{S}_\Pi$ is a point over the origin $O \in S_\Pi$ of the curves a_i, b_i, and $g \in \pi_1(S_\Pi)$. It also follows from the proof of the theorem in Section 6.5 that point $g\tilde{O}$ is connected to point $ga_i\tilde{O}$ (resp. $gb_i\tilde{O}$) by an edge of $\tilde{G}$ labeled a_i (resp. b_i). Thus, $\tilde{G}$ is to some extent a "picture" of $\pi_1(S_\Pi)$. We call $\tilde{G}$ the *Cayley graph* of $\pi_1(S_\Pi)$.

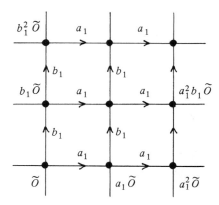

FIGURE 6.11.

When $n = 1$, the picture is clear (Figure 6.11) and it gives us a complete description of $\pi_1(S_\Pi)$. To show that $\tilde{G}$ gives a complete description of $\pi_1(S_\Pi)$ in general, we prove the following:

Proposition. $\tilde{G}$ *is connected and it partitions* $\tilde{S}_\Pi$ *into polygons isometric to* Π, *each with boundary labeled* $a_1 b_1 a_1^{-1} b_1^{-1} \dots a_n b_n a_n^{-1} b_n^{-1}$.

Proof. Since S_Π is a geometric surface, the identification map $\Gamma_\Pi : \Pi \to S_\Pi$ is a local isometry on the edges and corners of Π. In particular, the edges map to line segments of the same length in S_Π, meeting at the same angles. Then, because the covering map $\Gamma \cdot : \tilde{S}_\Pi \to S_\Pi$ is also a local isometry, the path $a_1 b_1 a_1^{-1} b_1^{-1} \dots a_n b_n a_n^{-1} b_n^{-1}$ in S_Π (the image of the boundary of Π under Γ_Π) lifts to a sequence of line segments in $\tilde{S}_\Pi$, labeled successively $a_1, b_1, a_1^{-1}, b_1^{-1}, \dots, a_n, b_n, a_n^{-1} b_n^{-1}$, of the same lengths as their namesakes on Π, and meeting at the same angles. These segments therefore enclose a polygon isometric to Π, which we may also call Π.

Thus, if A is a point of S_Π not on one of the curves a_i, b_i, any lift $\tilde{A}$ of A lies in a polygon isometric to Π. These lifts $\tilde{A}$ include *all* points in $\tilde{S}_\Pi - \tilde{G}$. Hence, $\tilde{G}$ partitions $\tilde{S}_\Pi$ into polygons isometric to Π, and each of them, being a lift of Π, has boundary labeled $a_1 b_1 a_1^{-1} b_1^{-1} \dots a_n b_n a_n^{-1} b_n^{-1}$.

To show that $\tilde{G}$ is connected, it will suffice to show that the connected union of polygons obtained by starting with Π, attaching adjacent polygons, then the polygons adjacent to these, etc., is the whole of $\tilde{S}_\Pi$. In turn, it suffices to show that any line segment originating in Π can be extended indefinitely through this union. But this is a consequence of the completeness of S_Π, which was proved in Section 5.7. $\qquad\square$

When the genus $n = 2$, the proposition gives a regular tessellation of $\mathbb{H}^2$ by octagons, shown in the $\mathbb{D}^2$-model (without edge labels) in Figure 6.12. Eight octagons meet at each vertex because the neighborhood of a vertex

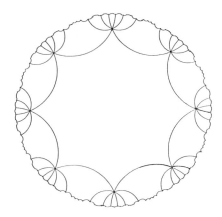

FIGURE 6.12.

on $\tilde{S}_\Pi$ is isometric to the neighborhood of the single vertex on S_Π, where the eight ends of a_1, b_1, a_2, b_2 meet.

To explain how $\tilde{G}$ captures the structure of $\pi_1(S_\Pi)$, we introduce the following terms. *Generators* of a group Γ are elements $g_1, g_2, \ldots \in \Gamma$ such that each $g \in \Gamma$ is expressible in the form

$$g = g_{j_1}^{m_1} \ldots g_{j_k}^{m_k}$$

(the right-hand side being called a *word* in the g_j). *Defining relations* of Γ are equations

$$r_1 = 1, \quad r_2 = 1, \quad \ldots$$

(where the r_i are words in the g_j) which are valid in Γ and which imply *all* equations valid in Γ.

Theorem. $\pi_1(S_\Pi)$ *is generated by* $a_1, b_1, \ldots, a_n, b_n$ *and has defining relation*

$$a_1 b_1 a_1^{-1} b_1^{-1} \ldots a_n b_n a_n^{-1} b_n^{-1} = 1.$$

Proof. Since $\tilde{G}$ is connected, any vertex $g\tilde{O}$ of $\tilde{G}$ ($g \in \pi_1(S_\Pi)$) is connected to $\tilde{O}$ by an edge path labeled $g_{j_1}^{m_1} \ldots g_{j_k}^{m_k}$, where each g_j is an a_i or b_i. This means

$$g = g_{j_1}^{m_1} \ldots g_{j_k}^{m_k}$$

by the proof of the theorem in Section 6.5, and hence the a_i, b_i are generators of $\pi_1(S_\Pi)$.

If r is a word in the a_i, b_i and $r = 1$ holds in $\pi_1(S_\Pi)$, then the path (with origin $\tilde{O}$, say) in $\tilde{G}$ labelled r is closed. This is because

$$r = 1 \text{ in } \pi_1(S_\Pi) \Leftrightarrow \text{path } r \text{ in } S_\Pi \text{ is homotopic to } O$$

$$\Leftrightarrow \text{path } \tilde{r} \text{ labelled } r \text{ in } \tilde{S}_\Pi \text{ is homotopic to } \tilde{O}$$

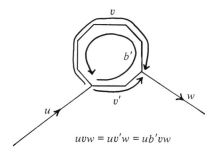

$$uvw = uv'w = ub'vw$$

FIGURE 6.13.

(by the homotopy lifting property)

$$\Leftrightarrow \text{path } \tilde{r} \text{ is closed}$$

($\Rightarrow$ because endpoints are fixed by homotopy, $\Leftarrow$ by the theorem in Section 6.3).

Now a closed path $\tilde{r}$ in $\tilde{G}$ can be decomposed into simple closed paths, each enclosing finitely many polygons. Hence, $\tilde{r}$ can be contracted to $\tilde{O}$ by finitely often pulling it across polygons, which amounts to applying relations

$$b' = \text{boundary path of } \Pi = 1 \text{ (Figure 6.13)},$$
$$uvw = uv'w = ub'vw$$

and removing "backtracking" (Figure 6.14), which amounts to cancellation of inverses

$$ua_i a_i^{-1} v = uv.$$

Since any boundary path b' of Π is a cyclic rearrangement of

$$b = a_1 b_1 a_1^{-1} b_1^{-1} \ldots a_n b_n a_n^{-1} b_n^{-1}$$

or of b^{-1}, in which case $b' = 1$ is a consequence of $b = 1$, it follows that all relations of $\pi_1(S_\Pi)$ are consequences of

$$a_1 b_1 a_1^{-1} b_1^{-1} \ldots a_n b_n a_n^{-1} b_n^{-1} = 1. \qquad \square$$

Exercises

6.6.1. If Π is the $4n$-gon with boundary $a_1 b_1 a_1^{-1} b_1^{-1} \ldots a_n b_n a_n^{-1} b_n^{-1}$, show that $4n$ edges meet at each vertex of $\tilde{G}$ on $\tilde{S}_\Pi$.

6.6.2. Consider the following topological model of the tessellation of $\tilde{S}_\Pi$ by the $4n$-gons $\Pi, \Pi', \Pi'', \ldots$.

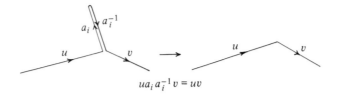

$$ua_i a_i^{-1} v = uv$$

FIGURE 6.14.

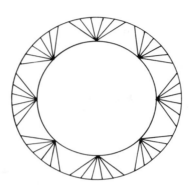

FIGURE 6.15.

Take a circle C_1, subdivided by $4n$ vertices, as the boundary of Π. Take a circle C_2 concentric with C_1, and $4n - 2$ edges from each vertex on C_1 to C_2, dividing the annulus between C_1 and C_2 into 3-gons and 4-gons. Figure 6.15 shows the case $n = 2$.

Now add extra vertices on C_2 to make each of these a $4n$-gon. Continue similarly with concentric circles C_3, $C_4, \ldots$ at each stage arranging that $4n$ $4n$-gons meet at each vertex. (Remark: In the $\mathbb{D}^2$-model, the successive layers of octagons actually come quite close to lying on concentric circles. Compare Figure 6.15 with Figure 6.12.)

Show by induction on m that a $4n$-gon lying between C_m and C_{m+1} has at least $4n - 3$ edges on C_{m+1}.

6.6.3. Deduce that a closed edge path in $\tilde{G}$, if it is nontrivial and has no backtracking, contains a segment consisting of at least $4n - 2$ consecutive edges in some face boundary.

6.6.4 (Dehn's algorithm). If $r = 1$ holds in $\pi_1(S_\Pi)$, show that the word r may be reduced to 1 by cancellation of adjacent inverses and replacement of subwords v corresponding to more than half a polygon boundary by shorter words (cf. Figure 6.13).

6.7 Fundamental Group and Genus

We now have an algebraic description of the fundamental group $\pi_1(S_n)$ of an orientable surface $S_\Pi = S_n$ of genus $n \geq 1$: generators $a_1, b_1, \ldots, a_n, b_n$; defining relation $a_1 b_1 a_1^{-1} b_1^{-1} \ldots a_n b_n a_n^{-1} b_n^{-1} = 1$. We abbreviate this by

$$\pi_1(S_n) = \langle a_1, b_1, \ldots, a_n, b_n \mid a_1 b_1 a_1^{-1} b_1^{-1} \ldots a_n b_n a_n^{-1} b_n^{-1} = 1 \rangle.$$

This description enables us to show, at last, that surfaces of different genus have nonisomorphic fundamental groups, and hence are nonhomeomorphic. The idea is to *abelianize* $\pi_1(S_n)$ by adding relations $gh = hg$ for all $g, h \in \pi_1(S_n)$. The resulting group is called $H_1(S_n)$ (the H standing for "homology" because this group is commonly obtained via homology theory, which we do not discuss in this book), and we have the following:

Theorem. *If $n \neq n'$, then $H_1(S_n) \not\cong H_1(S_{n'})$.*

Proof. The relations $gh = hg$ for all $g, h \in \pi_1(S_n)$ are consequences of the smaller set of relations

$$gh = hg \quad \text{for } g, h \in \{a_1, b_1, \ldots, a_n, b_n\} \tag{1}$$

since the a_i, b_i generate all g, h. The defining relation

$$a_1 b_1 a_1^{-1} b_1^{-1} \ldots a_n b_n a_n^{-1} b_n^{-1} = 1$$

of $\pi_1(S_n)$ is also a consequence of (1); hence, $H_1(S_n)$ is simply the group with $2n$ generators $a_1, b_1, \ldots, a_n, b_n$ and the relations (1), which say that any two of the generators commute.

Thus, any element of $H_1(S_n)$ has a unique normal form

$$a_1^{m_1} b_1^{m_2} \ldots a_n^{m_{2n-1}} b_n^{m_{2n}},$$

where $m_1, \ldots, m_{2n} \in \mathbb{Z}$ and hence $H_1(S_n) \cong \mathbb{Z}^{2n}$ via the correspondence

$$a_1^{m_1} b_1^{m_2} \ldots a_n^{m_{2n-1}} b_n^{m_{2n}} \leftrightarrow (m_1, m_2, \ldots, m_{2n}) \in \mathbb{Z}^{2n}.$$

But it is clear (e.g., by using linear algebra to show that any independent set of generators of $\mathbb{Z}^{2n}$ has $2n$ elements; see Exercises) that $\mathbb{Z}^{2n} \not\cong \mathbb{Z}^{2n'}$ when $n \neq n'$; hence $H_1(S_n) \not\cong H_1(S_{n'})$. □

Corollary. *S_n is not homeomorphic to $S_{n'}$ when $n \neq n'$.*

Proof. If S_n is homeomorphic to $S_{n'}$, then $\pi_1(S_n) \cong \pi_1(S_{n'})$ by the topological invariance of π_1 (Section 6.5). If $\pi_1(S_n) \cong \pi_1(S_{n'})$, then $H_1(S_n) \cong H_1(S_{n'})$ because the abelianizations of isomorphic groups are isomorphic (abelianization was defined by adding the relations $gh = hg$ for *all* g, h to make this obvious). Hence $n = n'$ by the theorem. □

Exercises

6.7.1. Show that $\mathbb{Z}^{2n}$ needs at least $2n$ generators, by a vector space dimension argument.

6.7.2. Show that $m > 2n$ elements of $\mathbb{Z}^{2n}$ are linearly dependent with rational (and hence, after multiplying through by a common denominator, integer) coefficients.

6.7.3. Conclude that any independent set of generators of $\mathbb{Z}^{2n}$ has $2n$ elements, and hence that $\mathbb{Z}^{2n} \not\cong \mathbb{Z}^{2n'}$ when $n \neq n'$.

6.8 Closed Geodesic Paths

In earlier chapters we have taken a line on a surface $S = \tilde{S}/\Gamma$ to be the *image* of a line on $\tilde{S}$ under the orbit map, not the map itself. Our experience with paths and homotopy suggests that we should refine this notion. For example, it would be desirable to distinguish the path p that winds once around the cylinder (Figure 6.16) from the path p^2 that winds around twice. As we know, the formal difference between p and p^2 is that they are maps of different line segments on $\tilde{S} = \mathbb{R}^2$, the line segment for p^2 being twice as long as the one for p.

We therefore define a *geodesic path* p on $S = \tilde{S}/\Gamma$ to be the restriction of the orbit map $\Gamma \cdot : \tilde{S} \to \tilde{S}/\Gamma$ to a line segment $\tilde{p}$ of $\tilde{S}$ whose image is a line on $\tilde{S}$ (i.e., the image of a line on $\tilde{S}$). As usual, p is *closed* if $\Gamma \cdot$ maps the endpoints of $\tilde{p}$ to the same point on S.

The reason for explicitly requiring the image to be a line is that the image of a line *segment*, even when closed, is not necessarily a line. The two ends may meet at a nonzero angle, to form a so-called *geodesic monogon*. The simplest examples occur on the pseudosphere (Figure 6.17). We take the pseudosphere $\mathbb{H}^2$ as $\mathbb{H}^2/\Gamma$, where Γ is the group generated by $z \mapsto 1 + z$, and consider a hyperbolic line segment $\tilde{p}$ between a point $\tilde{O} = -\frac{1}{2} + iy$ and $\tilde{O}' = \frac{1}{2} + iy$. Since $\tilde{O}, \tilde{O}'$ lie over the same point O of $\mathbb{H}^2/\Gamma$, the map p of $\tilde{p}$ is closed. But since $\tilde{p}$ is a circular arc, it meets its translate $\tilde{p}' = 1 + \tilde{p}$ at a nonzero angle at $\tilde{O}'$, and this is the angle at which the ends of p meet (because the neighborhood of O is isometric to the neighborhood of $\tilde{O}'$).

The presence of geodesic monogons (which also occur on compact sur-

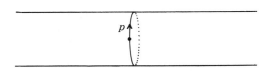

FIGURE 6.16.

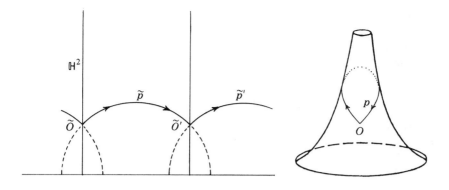

FIGURE 6.17.

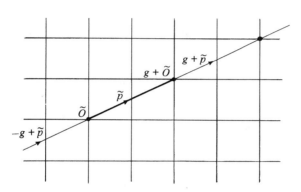

FIGURE 6.18.

faces—see Exercise 6.8.2 and Section 6.9) complicates the problem of classifying geodesic paths. For the remainder of this section we shall consider only the torus, on which there are no geodesic monogons and whose geodesic paths are elegantly classified in terms of homotopy. In Section 6.9, we will give a generalization of homotopy which solves the problem for other compact surfaces.

Let T be the torus $\mathbb{R}^2/\Gamma$, where Γ is generated by translations a, b, and let $\tilde{p}$ be a line segment between points $\tilde{O}$ and $g + \tilde{O}$ of $\mathbb{R}^2$, where $g \in \Gamma$. Figure 6.18 shows the case $g = 2a + b$. The segments $-g + \tilde{p}$, $\tilde{p}$, $g + \tilde{p}, \ldots$, which all map to the same p on T, obviously form a line in $\mathbb{R}^2$; hence, p is a geodesic path. This leads to the following:

Theorem. *Each nontrivial homotopy class with origin O on T includes a unique geodesic path. Conversely, each closed geodesic path on T is homotopically nontrivial.*

Proof. Let p be a closed path on T with origin O, let $\tilde{O}$ be a point of $\mathbb{R}^2$ over O, and consider the lift $\tilde{p}$ of p with origin $\tilde{O}$. If p is homotopically nontrivial, then the terminus $\tilde{O}'$ of $\tilde{p}$ is not $\tilde{O}$. Hence, there is a unique line segment $\alpha(\tilde{p})$ from $\tilde{O}$ to $\tilde{O}'$. By Section 6.3, $\alpha(\tilde{p})$ is in the same homotopy class as $\tilde{p}$, and, hence, its projection $\Gamma\alpha(\tilde{p}) = \alpha(p)$ is a closed geodesic in the homotopy class of p.

Conversely, suppose q is a closed geodesic with origin O on T. Let $\tilde{O}$ be a point of $\mathbb{R}^2$ over O and consider the lift $\tilde{q}$ of q with origin $\tilde{O}$. Since q is a geodesic path, $\tilde{q}$ is a line segment in $\mathbb{R}^2$, which means, in particular, that its terminus is a point $\tilde{O}' \not\cong \tilde{O}$. It follows that $\tilde{q}$, and hence q, is not homotopically trivial. □

Exercises

6.8.1. Show that there are no closed geodesic paths on the pseudosphere.

6.8.2. Show that there are geodesic monogons on the Klein bottle.

6.9   Classification of Closed Geodesic Paths

When we consider compact orientable surfaces of genus $n > 1$, geodesic monogons arise immediately. Let S_2 be the (hyperbolic) surface of genus 2 obtained from a regular octagon Π with angles $\pi/4$ and boundary path $a_1 b_1 a_1^{-1} b_1^{-1} a_2 b_2 a_2^{-1} b_2^{-1}$. Pasting together the sides of Π as the boundary prescribes, we see that the corners of Π come together to form a vertex O on S_2 with edges in the cyclic order shown in Figure 6.19. The outgoing and incoming edge at O labeled a_1 are the two ends of the line segment a_1 on $S_2 = S_{\Pi}$. Thus, a_1 is a geodesic monogon which meets itself at an angle $\pi/2$. It is no use looking for a closed geodesic homotopic to a_1 because any a_1^* homotopic to a_1 lifts to an a_1^* from $\tilde{O}$ to $\tilde{O}'$ on $\mathbb{H}^2$ (Figure 6.19, which actually shows $\mathbb{D}^2$) and the lift of a_1 is already the unique line segment from $\tilde{O}$ to $\tilde{O}'$.

To find a closed geodesic "topologically related" to a_1, we relax the notion of deformation to allow movement of the origin O. This notion of deformation, called *free homotopy*, is defined the same as homotopy but without the requirement that the endpoints be fixed (though, of course, for a closed curve, the endpoints have to be identical at all times, and, hence, the endpoints of its lift have to be in the same Γ-orbit).

To see the closed geodesic to which a closed curve g on S_n $(n \geq 2)$ can be deformed, one envisages the lifts of g linking the points $\ldots g^{-1}\tilde{O}$, $\tilde{O}$, $g\tilde{O}$, $g^2\tilde{O}, \ldots$ on $\tilde{S}_n$, shown in the $\mathbb{D}^2$-model in Figure 6.20. The following proposition justifies the general appearance of Figure 6.20:

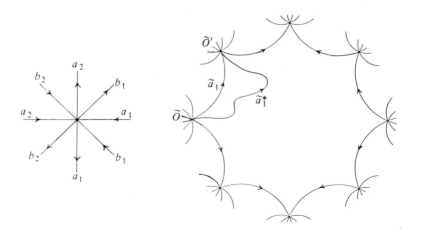

FIGURE 6.19.

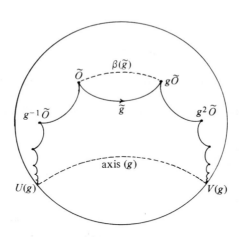

FIGURE 6.20.

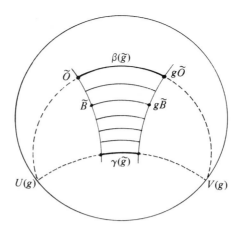

FIGURE 6.21.

Proposition. *An element $g \neq 1$ of $\pi_1(S_n)$ is a translation, and the points $\ldots g^{-1}\tilde{O}, \tilde{O}, g\tilde{O}, g^2\tilde{O}, \ldots$ lie on an equidistant curve of the axis of g.*

Proof. It follows from the description of the Cayley graph $\tilde{G}$ in Section 6.6 that the isometry $g \in \Gamma$ of $\mathbb{D}^2$ moves each point of $\mathbb{D}^2$ through a distance at least as great as the side length of a polygon. Hence, the isometry g is a translation by the classification of hyperbolic isometries (Section 4.6).

Now $\mathbb{D}^2$ is partitioned into invariant curves of g which are the equidistant curves of the axis of g (Section 4.5). The set $\{\ldots g^{-1}\tilde{O}, \tilde{O}, g\tilde{O}, g^2\tilde{O}, \ldots\}$ is mapped onto itself by g; hence, it lies along an equidistant curve. $\qquad\square$

This proposition suggests that, if we allow the origin to move, then the path g on S_n can be deformed to a closed geodesic covered by the axis of g. A closer look at the situation gives the following:

Theorem. *Each nontrivial free homotopy class on S_n has a unique geodesic representative. Conversely, each closed geodesic path represents a nontrivial free homotopy class.*

Proof. First we deform the lift $\tilde{g}$ of g on $\mathbb{D}^2$ to the segment $\beta(\tilde{g})$ of the equidistant curve through $\tilde{O}$ and $g\tilde{O}$ (Figure 6.20). This is possible because any two paths on $\mathbb{D}^2$ with the same endpoints are homotopic (Section 6.3), and the homotopy projects to a homotopy on S_n, between the path g and a path $\beta(g)$.

Now deform $\beta(\tilde{g})$ through segments of equidistant curves to a segment $\gamma(\tilde{g})$ of axis(g), in such a way that its ends remain on the perpendiculars to axis(g) through $\tilde{O}$ and $g\tilde{O}$ (Figure 6.21). This means that the ends of each

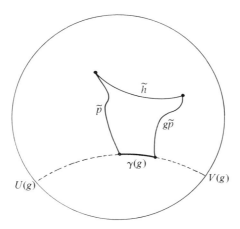

FIGURE 6.22.

intervening segment are of the form $\tilde{B}$, $g\tilde{B}$, so the free homotopy on $\mathbb{D}^2$ projects to a free homotopy on S_n, from $\beta(g)$ to the closed geodesic $\gamma(g)$ that is the projection of $\gamma(\tilde{g})$.

The geodesic path $\gamma(g)$ is therefore a representative of the free homotopy class of g. To see that it is unique, consider any curve h freely homotopic to $\gamma(g)$, and let p be the path from the origin of $\gamma(g)$ to the origin of h traced during the homotopy. By lifting the free homotopy (by the same method used to lift homotopies in Section 6.5), we get a lift $\tilde{h}$ of h and lifts $\tilde{p}$, $g\tilde{p}$ of p as shown in Figure 6.22. Applying powers of the translation g to this quadrilateral we get an "infinite ladder" like that shown in Figure 6.23. Since the various lifts $\ldots g^{-1}\tilde{h}, \tilde{h}, g\tilde{h}, \ldots$ all lie within a constant hyperbolic distance of axis(g), their *euclidean* distance from axis(g) tends to 0 at the ends of axis(g). That is, the chain $\ldots g^{-1}\tilde{h}, \tilde{h}, g\tilde{h}, \ldots$ covering h has the same endpoints $U(g)$, $V(g)$ on $\partial\mathbb{D}^2$ as axis(g). Since axis(g) is the unique hyperbolic line through $U(g)$, $V(g)$, the chain is not a line, and hence h is not a closed geodesic.

This completes the proof that each nontrivial free homotopy class has a unique geodesic representative.

Conversely we find, as in the case of the torus, that any closed geodesic q on S_n lifts to a line segment $\tilde{q}$ with distinct endpoints, say $\tilde{O}$ and $g\tilde{O}$, with $g \neq 1$. Under a free homotopy the endpoints of the lift remain of the form $\tilde{B}$, $g\tilde{B}$, hence distinct. Hence, the free homotopy class of q is nontrivial. $\square$

Exercises

6.9.1. Find a closed geodesic on S_2 which has the form of a figure 8.

6.9.2. Construct the neighborhood of the vertex on the S_3 which results

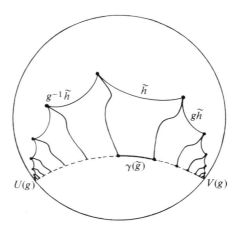

FIGURE 6.23.

from a regular 12-gon with boundary $a_1 b_1 a_1^{-1} b_1^{-1} a_2 b_2 a_2^{-1} b_2^{-1} a_3 b_3 a_3^{-1} b_3^{-1}$. Hence, find a closed geodesic on S_3 which has the form of a three-leaved clover.

6.9.3. Investigate the closed geodesics on the Klein bottle.

6.10   Discussion

The fundamental group is one of Poincaré's great contributions to mathematics (Poincaré [1892, 1895]). By associating a topologically invariant group with each manifold, he gave the first general method for proving the topological distinctness of different manifolds. The idea of homotopy itself is much older and can be traced back to the ideas of Gauss on integration in the complex plane. In his letter to Bessel, Gauss [1811] considers the integration of $1/x$ in $\mathbb{C} - \{0\}$, essentially observing that homotopic paths give the same integral and that the "function" $\log z = \int_1^z \frac{dx}{x}$ has multiple values corresponding to the multiplicity of *non*homotopic paths in $\mathbb{C} - \{0\}$ from 1 to z.

With a little hindsight, one can see that resolving the multivaluedness of an integral amounts to constructing the universal covering as a space of homotopy classes. In particular, resolving the multivaluedness of $\int_1^z \frac{dx}{x}$ leads to a proper understanding of the complex exponential function and shows why it defines a covering of $\mathbb{C} - \{0\}$ by $\mathbb{C}$ (cf. Section 2.10).

As Gauss observed, $\int_1^z \frac{dx}{x}$ depends on the homotopy class, in $\mathbb{C} - \{0\}$, of the path p of integration from 1 to z. This is because $\int_p \frac{dx}{x}$ does not change as long as p does not pass through the point 0 where $1/x$ becomes infinite. Conversely, the integrals of $1/x$ along nonhomotopic paths, such as p_1 and

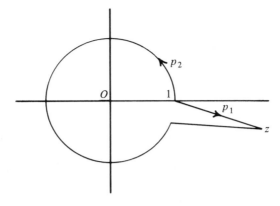

FIGURE 6.24.

p_2 in Figure 6.24, are different. They differ by $2in\pi$, where n is the number of anticlockwise windings about 0 that p_2 has in excess of p_1. This can be seen by deforming the paths so that they differ only by n circuits of the unit circle c (as p_1, p_2 in Figure 6.24 are already close to doing), and then calculating $\int_c \frac{dx}{x} = 2i\pi$.

Thus if p is a path in $\mathbb{C} - \{0\}$ from 1 to z, then $\int_p \frac{dx}{x}$ is actually a function of the homotopy class of p. This homotopy class is determined by two pieces of data: the endpoint z and its "tail", the winding number n. The winding numbers n correspond to elements of $\pi_1(\mathbb{C} - \{0\})$, and indeed $\pi_1(\mathbb{C} - \{0\}) \cong \mathbb{Z}$ because a path with winding number m, followed by a path with winding number n, is a path with winding number $m + n$.

Correspondingly, the value of $\int_p \frac{dx}{x} = \int_1^z \frac{dx}{x}$ is the sum of a *principal value* Log z, with imaginary part between $-\pi$ and π, and a winding number times $2i\pi$. These values fill $\mathbb{C}$ (cf. Figure 2.21), so the inverse function exp maps $\mathbb{C}$ onto $\mathbb{C} - \{0\}$, with the number $\log z = \text{Log } z + 2in\pi \in \mathbb{C}$, corresponding to the number $z \in \mathbb{C} - \{0\}$ at the end of a "tail" with winding number n, being mapped to $z = \exp(\log z)$.

This concrete example makes a very natural connection between coverings and homotopy classes; however, it is not the whole story. A connection still exists when the analysis, the geometry, and most of the topology is stripped away. The construction of the fundamental group applies to *any* reasonable space S and shows that S is homeomorphic to $\tilde{S}/\pi_1(S)$. (This is actually an almost trivial piece of "abstract nonsense". $\tilde{S}$ is the space of points of S with tails. The group $\pi_1(S)$ acts on $\tilde{S}$ as follows. Each $g \in \pi_1(S)$ sends point P with tail t to point P with tail tg. The π_1-orbit of P with tail t is therefore the point P *with all possible tails*, hence the π_1-orbits—the points of $\tilde{S}/\pi_1(S)$—correspond to the points P of S.)

Thus, the relation between the universal covering and the homotopy classes is meaningful at the purely topological level. In fact, one can go on to construct not just the universal covering $\tilde{S}$, for which $\pi_1(\tilde{S}) = \{1\}$, but a covering S^* whose π_1 is any subgroup of $\pi_1(S)$. See e.g., Massey [1967, p. 173]. Conversely, each covering S^* of S is such that $\pi_1(S^*)$ is a subgroup of $\pi_1(S)$. So perhaps coverings are really about the relation between groups and subgroups?

The group-theoretic view of coverings can be presented most convincingly by reducing the topology to combinatorics, as was done by the discoverer of the covering–subgroup correspondence, Reidemeister [1928, 1932]. The topology of a compact surface S, for example, is captured by a finite description which lists the vertices and edges of a polygon Π and their identifications. A covering S^* of S is a tessellation paved by copies of Π which, as we have seen in Section 6.6, can be interpreted as a picture of a group. One can also redefine the concept of homotopy in a combinatorial manner and obtain a combinatorial fundamental group isomorphic to the topological one.

This methodology seems the best way to solve many problems in the so-called *combinatorial group theory*—the theory of groups defined by generators and relations—as one may see demonstrated in Zieschang, Vogt, and Coldewey [1980] and D.E. Cohen [1989]. A basic problem in combinatorial group theory is the *word problem*, solved combinatorially for the fundamental groups of surfaces in Exercises 6.6.1–6.6.4. Another is the *conjugacy problem*: to decide for arbitrary elements g_1, g_2 in a group G whether $g_1 = h^{-1}g_2h$ for some $h \in G$. When $G = \pi_1(S)$, the conjugacy problem is equivalent to the problem of deciding whether curves g_1, g_2 are freely homotopic in S, which we solved using geodesics in Section 6.9. (This solution is based on the work of Poincaré [1904].) There is also a purely combinatorial solution, due to Dehn [1912].

It nevertheless remains true that some purely group-theoretic or topological problems have been solved only with the help of geometry. And, of course, geometry will necessarily be involved when the problems themselves have geometric content, as is the case in the next two chapters.

7

Planar and Spherical Tessellations

7.1 Symmetric Tessellations

By a *tessellation* T we mean a division of a geometric surface into non-overlapping congruent polygons, called *tiles*. T is *symmetric* if it "looks the same from the viewpoint of any tile", i.e., if any tile Π_1 can be mapped onto any tile Π_2 by an isometry which maps the whole of T onto itself (faces onto faces and edges onto edges). The isometries of T onto itself are called *symmetries* of T, and they form a group called the *symmetry group* of T. Thus, we are defining T to be symmetric if its symmetry group contains enough elements to map any tile onto any other tile.

An example of a *non*symmetric tessellation T is shown in Figure 7.1. The symmetries of this T are precisely the translations of $\mathbb{R}^2$ with integral x and y components. Thus, even though all the tiles are congruent, there is no symmetry of T which maps Π_1 onto Π_2. T "looks different" from the viewpoints of Π_1 and Π_2.

The most familiar symmetric tessellations are the tessellations of $\mathbb{S}^2$ corresponding to the regular polyhedra (Section 3.9) and the tessellations of $\mathbb{R}^2$ based on the equilateral triangle, regular hexagon, and square (Figure 7.2). The simultaneous view of the triangle and hexagon (dotted) tessellations shows that they have the same symmetry group.

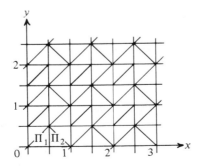

FIGURE 7.1.

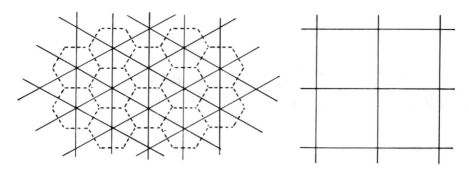

FIGURE 7.2.

FIGURE 7.3.

In each of these tessellations there are several symmetries which map a given tile onto itself. For example, all eight symmetries of the square are symmetries of the square tessellation. Because of this, the tessellation obtained by subdividing each square into eight triangles (Figure 7.3) has the same symmetry group as the square tessellation. The same idea gives a *common* subdivision of both the equilateral triangle and hexagon tessellations (Figure 7.4), and with the same symmetry group as both. Each triangle is divided into 6 and each hexagon into 12.

The subdivided tessellation, whose tiles we will call *subtiles* to distinguish them from the original tiles, gives a better picture of the symmetry group Γ. Since the subtiles have been chosen within an original tile so that exactly one $g \in \Gamma$ maps a given one of them onto another, the same is true for any two subtiles in the tessellation. Thus, if we choose a fixed subtile Π to correspond to $1 \in \Gamma$, any other tile $g\Pi$ corresponds to a unique $g \in \Gamma$. In other words, Π is a *fundamental region* for Γ because it includes a representative from each Γ-orbit, and each Γ-orbit is represented at most once in the interior of Π.

A useful enhancement of these tessellations is obtained by coloring the subtiles alternately black and white (Figure 7.5). In each case, the coloring is emphasised on one copy of the original tile. The symmetries of the original tile (and hence of the whole tessellation) which preserve the coloring are precisely the *orientation-preserving* symmetries (four for the square, three for the triangle). The union of a black subtile and a white subtile is

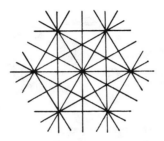

FIGURE 7.4.

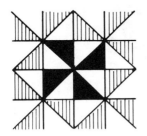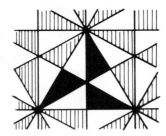

FIGURE 7.5.

a fundamental region for the subgroup $\Gamma^+ \subset \Gamma$ of orientation-preserving symmetries. We will give a general explanation of this method of passing from a fundamental region for Γ to a fundamental region for Γ^+ in the next section.

The tessellations of $\mathbb{S}^2$ which represent the polyhedral groups can likewise be given this "checkerboard" coloring to expose the orientation-preserving subgroups. Figure 7.6 shows the tessellation for the icosahedral group.

We have previously seen examples of tessellations by images of a fundamental region (e.g., Sections 5.3 and 6.6), and we have used them as "pictures" of the corresponding groups Γ. The present situation is similar except that the action of Γ need no longer be fixed point free. One of the main objectives of this chapter is to arrive at a geometric understanding of groups with fixed points. This goes hand in hand with an understanding of tessellations.

A basic question is: which polygons Π occur as fundamental regions for discontinuous groups? We shall begin to answer this question in the next section by finding necessary conditions for a compact polygon Π to be a fundamental region. These conditions will be shown to be sufficient in Section 7.4. Our emphasis will be on compact polygons throughout (reasons for this will also be given in Section 7.4), though noncompact examples will occasionally be mentioned.

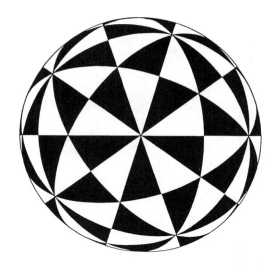

FIGURE 7.6.

Exercises

7.1.1. Consider the subtile Π of the square shown in Figure 7.7 as a fundamental region for the group Γ^+ of orientation-preserving symmetries of the square tessellation. Show that Γ^+ is generated by

g = clockwise rotation through $\pi/2$ about the center of the square, and

h = rotation through π about the midpoint of an edge of the square.

7.1.2. Deduce that Γ^+ has the Cayley graph shown in Figure 7.8. (Subtiles are indicated by dotted lines. The solid lines with arrows are the g-edges, those without are the h-edges.)

7.1.3. Conclude that Γ^+ has the description (called a *presentation*)

$$\langle g, h \mid g^4 = h^2 = (gh)^4 = 1 \rangle.$$

7.1.4. Show similarly that the group of orientation-preserving symmetries of the equilateral triangle tessellation has presentation

$$\langle g, h \mid g^3 = h^2 = (gh)^6 = 1 \rangle$$

and that the groups of orientation-preserving symmetries of the icosahedron, octahedron, and tetrahedron are, respectively,

$$\langle g, h \mid g^3 = h^2 = (gh)^5 = 1 \rangle,$$
$$\langle g, h \mid g^3 = h^2 = (gh)^4 = 1 \rangle,$$
$$\langle g, h \mid g^3 = h^2 = (gh)^3 = 1 \rangle.$$

FIGURE 7.7.

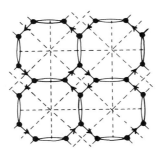

FIGURE 7.8.

7.1.5. Show that the even permutations

$$g = (2,3,4), \quad h = (1,2)(3,4)$$

satisfy

$$g^3 = h^2 = (gh)^3 = 1$$

and that g, h generate all even permutations of four things. Deduce, using the fact that the group A_4 of even permutations of four things has the same number of elements as $\langle g, h \mid g^3 = h^2 = (gh)^3 = 1 \rangle$, that these two groups are isomorphic.

7.1.6. Using $g = (2,3,4)$, $h = (1,4)$ show that

$$S_4 \cong \langle g, h \mid g^3 = h^2 = (gh)^4 = 1 \rangle$$

and using $g = (1,3,5)$, $h = (1,2)(3,4)$ show that

$$A_5 \cong \langle g, h \mid g^3 = h^2 = (gh)^5 = 1 \rangle.$$

7.2 Conditions for a Polygon to Be a Fundamental Region

To be able to state these conditions we subdivide the boundary of a polygonal fundamental region Π by finitely many points, called *vertices*, so that the boundary segments between vertices, called *sides*, coincide with sides of

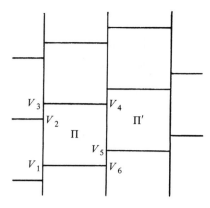

FIGURE 7.9.

polygons adjacent to Π. For example, in the symmetric tessellation shown in Figure 7.9 the fundamental polygon Π is a square, but it has six vertices $V_1, \ldots, V_6$, and consequently six sides. When each square $\Pi' = g\Pi$ is given the vertices $gV_1, \ldots, gV_6$ then the sides of Π coincide with sides of polygons adjacent to Π.

Any fundamental polygon Π for a discontinuous group Γ can be given sides which coincide with sides of its neighbors in this way, and if Π is compact, then the number of sides must be finite, otherwise Π has infinitely many neighbors and, hence, Γ is not discontinuous, by the argument of Section 5.8.

Theorem. *If Γ is a group acting discontinuously on $\mathbb{S}^2$, $\mathbb{R}^2$, or $\mathbb{H}^2$ with fundamental polygon Π, and if Γ includes an orientation-reversing element b, then $\Pi \cup b\Pi$ is a fundamental region for the subgroup Γ^+ of orientation-preserving elements. Moreover, b can be chosen so that $\Pi \cup b\Pi$ is a polygon.*

Proof. The set $\Gamma^+ b = \{gb \mid g \in \Gamma^+\}$ is the coset of orientation-reversing elements and Γ is the disjoint union of Γ^+ and $\Gamma^+ b$. Disjointness means we can color the Γ^+-images of Π white and the $\Gamma^+ b$-images of Π black. And since the union of Γ^+ and $\Gamma^+ b$ is Γ, the Γ^+-images of Π, $b\Pi$ include each Γ-image of Π exactly once. That is, $\Pi \cup b\Pi$ is a fundamental region for Γ^+.

Since the surface on which Γ acts is filled with black and white tiles, there must be adjacent black and white tiles, and hence a black tile Π^* adjacent to Π (i.e., having an edge in common with Π) by symmetry. We take b to be the element of Γ which maps Π onto Π^*.

If $\Pi \cup b\Pi$ is not a polygon, i.e., if its boundary is not a simple curve, then by the polygonal Jordan curve theorem it encircles a region R of the surface (Figure 7.10). Since the Γ^+-images of $\Pi \cup b\Pi$ fill the surface, one of them must lie in R. (For if an image $\Pi' \cup b\Pi'$ lies only partly in R, it must be pinched at (say) X, hence so too is one of Π', $b\Pi'$ because Π',

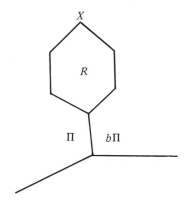

FIGURE 7.10.

$b\Pi'$ have more than one point in common. This contradicts the fact that Π', $b\Pi'$ are polygons.) But an image $\Pi' \cup b\Pi'$ in R will enclose another region R', containing another image $\Pi'' \cup b\Pi''$, and so on ad infinitum. This contradicts the discontinuity of Γ^+.   □

Remarks. (1) The process of "doubling" Π to $\Pi \cup b\Pi$ used in this theorem allows us to confine attention to the fundamental polygons for orientation-preserving groups Γ^+. It says that a fundamental polygon for a group Γ with orientation-reversing elements is "half" a fundamental polygon for the orientation-preserving subgroup Γ^+.

(2) A similar argument shows that if Γ' is a subgroup of Γ with distinct cosets $\Gamma'g_1$, $\Gamma'g_2, \ldots$ and if Π is a fundamental polygon for Γ, then $g_1\Pi \cup g_2\Pi\cup \ldots$ is a fundamental region for Γ'. The "coloring" argument used above generalizes to show that if there are only finitely many cosets $\Gamma'g_1, \ldots, \Gamma'g_n$ then $g_1, \ldots, g_n$ may be chosen so that the fundamental region $g_1\Pi\cup \ldots \cup g_n\Pi$ is a polygon. Namely, suppose that all tiles in $\Gamma'g_i\Pi$ are given "color g_i". Since the colors fill the surface, it must be possible to connect any color to any other by a chain of adjacent tiles. We can therefore find a union of finitely many tiles, each sharing an edge with another, which includes all colors. Also, by symmetry, we can eliminate a second occurrence of any color because the same configuration of colors occurs around the first occurrence. In this way, we reduce to a union of n cells, each of a different color, and each sharing an edge with another. This union is therefore a fundamental region, and the Jordan curve argument shows it to be a polygon.

The construction of a fundamental polygon $\Pi' = g_1\Pi \cup \ldots \cup g_n\Pi$ for a subgroup Γ' of finite index in Γ will be important in Chapter 8. We now return to our immediate goal, the statement of geometric conditions for Π to be a fundamental polygon.

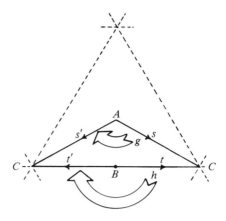

FIGURE 7.11.

Side and Angle Conditions. *If a compact polygon* Π *is a fundamental region for a group* Γ *of orientation-preserving isometries of* $\mathbb{S}^2$, $\mathbb{R}^2$, *or* $\mathbb{H}^2$ *then*

(i) *for each side* s *of* Π *there is exactly one other side* s' *of* Π *of the form* $s' = gs$ *with* $g \in \Gamma$ *(the elements* g *are called side-pairing transformations of* Π), *and*

(ii) *if each side* s *is identified with the corresponding* s', *then each set of vertices identified as a result corresponds to a set of corners of* Π *with angle sum* $2\pi/p$ *for some* $p \in \mathbb{Z}$ *(i.e., the corners sum to an "aliquot part" of* 2π).

Example. Figure 7.11 shows a fundamental polygon Π for the group Γ of orientation-preserving symmetries of the equilateral triangle tessellation. Also shown are the generators g, h of Γ (in the notation of Exercise 7.1.4), the sides s, s' paired by g, and the sides t, t' paired by h. Finally, the vertex cycles are labeled A (a singleton, with angle $2\pi/3$), B (also a singleton, with angle $\pi = 2\pi/2$) and C (a pair, with angle sum $\pi/6 + \pi/6 = 2\pi/6$).

Proof That the Conditions Hold. (i) Let $g_1\Pi, \ldots, g_k\Pi$ be the tiles which share sides with Π. Let s be a shared side of Π and $g_i\Pi$. Then $s' = g_i^{-1}s$ is a shared side of $g_i^{-1}\Pi$ and $g_i^{-1}g_i\Pi = \Pi$, hence also a side of Π. If $s' = s$, then $g_i^{-1} = g_i$, and because g_i is orientation preserving, this means g_i is a rotation through π about the midpoint of s (by the classification of isometries). If g_i is such a rotation, we declare the midpoint of s to be a new vertex, dividing s into sides s_1 and s_2. Then g_i maps s_1 onto s_2 and s_2 onto s_1; hence, in all cases a side s of Π is mapped by some $g \in \Gamma$ onto another side s' of Π. There is only one s' for each s, for if $s_1' = g_1s$ and

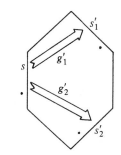

FIGURE 7.12.

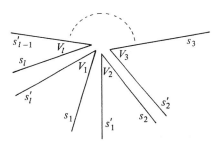

FIGURE 7.13.

$s_2' = g_2 s$ are both sides of Π, there are interior points of Π near s_1' and s_2' in the same Γ-orbit (Figure 7.12), contrary to the definition of fundamental region.

(ii) Suppose $\{V_1, \ldots, V_l\}$ is a vertex cycle of Π and suppose, in fact, that $\ldots V_l, V_1, V_2, \ldots, V_l, V_1, \ldots$ is the cyclic order of the corners at $V_1, \ldots, V_l$ induced by the side-pairing transformations (Figure 7.13). Then if $V = gV_j$ is an image of any such vertex under $g \in \Gamma$, the corners of tiles which meet at V are just images of the corners at $V_1, \ldots, V_l$, in the same cyclic order. The angle 2π at V is therefore an integer multiple of the angle sum of the corners at $V_1, \ldots, V_l$. $\square$

It seems plausible that polygon Π satisfying the side and angle conditions can always be used to tile the surface ($\mathbb{S}^2$, $\mathbb{R}^2$, or $\mathbb{H}^2$) to which it belongs, and indeed we shall prove this in Section 7.4. However, the assumption of compactness is essential because noncompact "polygons" may fail to fill the whole surface (see Exercises 7.2.1 and 7.2.2). Even for compact polygons there are some interesting difficulties, which can best be appreciated by looking first at the case of triangles. We do this in the next section.

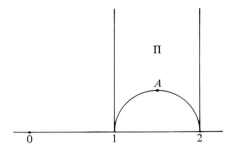

FIGURE 7.14.

Exercises

7.2.1. Consider the "polygon" $\Pi' \subset \mathbb{H}^2$ bounded by $\mathrm{Re}\,z = 1$ and $\mathrm{Re}\,z = 2$, and let Γ' be the group generated by $g(z) = 2z$. Show that $g\Pi'$ is a "polygon" adjacent to Π' but that the polygons $g^n\Pi'$, for $n \in \mathbb{Z}$, do not fill $\mathbb{H}^2$.

7.2.2. The polygon Π' in Exercise 7.2.1 may be considered to fail the side-pairing condition because only two of its three "sides" are paired. Consider therefore the polygon Π shown in Figure 7.14, and the rotation h through π about the point A.

Show that the images of Π under the group Γ generated by g and h also do not fill $\mathbb{H}^2$, even though g and h are side-pairing transformations for Π.

7.3   The Triangle Tessellations

The simplest polygons that satisfy the side and angle conditions are quadrilaterals obtained by doubling triangles. Doubling a triangle with angles π/p, π/q, π/r gives the quadrilateral shown in Figure 7.15. The side pairing is as shown, and the angle condition is satisfied just in case p, q, r are integers.

The symmetric tessellation based on the quadrilateral, if it exists, comes from that generated by reflections in the sides of the triangle (cf. the theorem in Section 7.2). The question of existence is settled by arguments which depend on the value of $\frac{1}{p} + \frac{1}{q} + \frac{1}{r}$. When

$$\frac{1}{p} + \frac{1}{q} + \frac{1}{r} > 1,$$

then all possibilities are realized by familiar tessellations of $\mathbb{S}^2$. When

$$\frac{1}{p} + \frac{1}{q} + \frac{1}{r} = 1,$$

then all possibilities are realized by familiar tessellations of $\mathbb{R}^2$. In each of

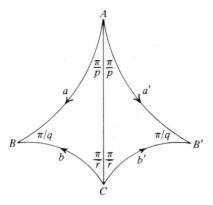

FIGURE 7.15.

these cases, the possibilities are easy to survey. When

$$\frac{1}{p} + \frac{1}{q} + \frac{1}{r} < 1,$$

there are infinitely many possibilities. In fact, they can all be realized in $\mathbb{H}^2$, though this is not obvious.

It is customary to call the triangle with angles π/p, π/q, π/r a (p, q, r) triangle. When

$$\frac{1}{p} + \frac{1}{q} + \frac{1}{r} \geq 1,$$

the tessellations by (p, q, r) triangles are subdivisions of the regular polyhedra (if we include "dihedra" among the polyhedra, cf. Section 3.8) and the regular tessellations of $\mathbb{R}^2$. For example, the $(2, 3, 5)$ triangle tessellation results from dividing each equilateral triangle of the icosahedral tessellation—which has angles $2\pi/5$—into six parts (Figure 7.16, and see also Figure 7.6). We summarize the results as follows:

p	q	r	Tessellation of $\mathbb{S}^2$
2	2	$n \geq 2$	Dihedron
2	3	3	Tetrahedron
2	3	4	Octahedron (or cube)
2	3	5	Icosahedron (or dodecahedron)

p	q	r	Tessellation of $\mathbb{R}^2$
3	3	3	Equilateral triangle (not subdivided)
2	4	4	Square
2	3	6	Equilateral triangle (or regular hexagon)

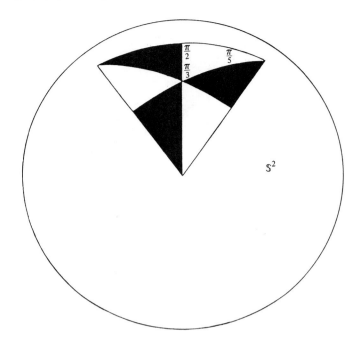

FIGURE 7.16.

All other values of p, q, $r \geq 2$ give $\frac{1}{p} + \frac{1}{q} + \frac{1}{r} < 1$ and, hence, an angle sum $< \pi$, so the (p, q, r) triangle must be hyperbolic. Proving that all these possibilities can be realized is, in fact, as difficult as proving that an *arbitrary* polygon satisfying the side and angle conditions can serve as a fundamental region; hence, we will postpone the proof until we deal with the general case. Instead we shall look at two examples which help to expose the difficulties involved.

Example 1. The $(4,4,4)$ triangle.

Sixteen $(4, 4, 4)$ triangles can be assembled to form a regular octagon in $\mathbb{H}^2$ with corner angles $\pi/4$ (Figure 7.17, which is from Burnside [1911, p. 395], shows this octagon in the $\mathbb{D}^2$-model). It follows that if we identify sides of the octagon so as to make a genus 2 surface with one vertex, then the angle sum at the vertex will be 2π, and hence S will be a hyperbolic surface by Section 5.5. We can then lift the tessellation of S to its universal cover, which is $\mathbb{H}^2$ by the Killing–Hopf theorem, giving an obviously symmetric tessellation of $\mathbb{H}^2$ by $(4, 4, 4)$ triangles.

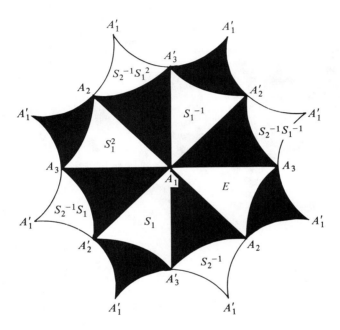

FIGURE 7.17. From Burnside: *The Theory of Groups of Finite Order*, p. 395, with permission from Dover Publications, Inc.

FIGURE 7.18.

Example 2. The $(2, 3, 7)$ triangle.

Klein [1879] made the astonishing discovery that there is a regular 14-gon with corner angles $2\pi/7$ formed by 336 copies of the $(2, 3, 7)$ triangle (Figure 7.18). If side $2i + 1$ of this 14-gon is identified with side $2i + 6$ (mod 14), the result turns out to be a genus 3 surface S with two vertices. The two vertices of S correspond to the sets of even and odd vertices of the 14-gon; hence, they each have angle sum 2π. Thus, we again have a hyperbolic surface, and we obtain the desired tessellation of $\mathbb{H}^2$ by lifting the tessellation of S to its universal cover.

Knowing now that the tessellation of $\mathbb{H}^2$ exists, we can see that it is symmetric because it is generated by reflections in the sides of the $(2, 3, 7)$ triangle. (The mere *appearance* of regularity in a tessellation should not be trusted, incidentally. See Exercise 7.3.3).

In these two examples we have used a compact surface as a shortcut to $\mathbb{H}^2$. Instead of filling $\mathbb{H}^2$ with infinitely many triangles, we have filled a compact surface with finitely many, then used the covering of the com-

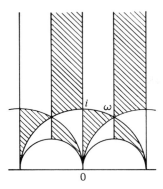

FIGURE 7.19.

pact surface by $\mathbb{H}^2$ guaranteed by the Killing–Hopf theorem. However, we may have been lucky to find these compact surfaces—indeed, the one for the $(2, 3, 7)$ triangle seems quite miraculous—and it is not clear that this approach will always work.

So, instead, we shall approach the general problem head on in the next section, by direct construction of an infinite tessellation. The Killing–Hopf theorem again plays an important role, but not the tessellation of compact surfaces. The latter is indeed a more difficult problem, to which we will return in Chapter 8.

Before proceeding to these results for compact polygons, it is worth looking at the famous modular tessellation (Section 5.9), whose tiles can be regarded as $(2, 3, \infty)$ triangles. We take the hyperbolic triangle with vertices at i, $\omega = \frac{1}{2} + \frac{\sqrt{3}}{2}i$ and ∞ in the half-plane model $\mathbb{H}^2$ and double it by reflection in the i axis (Figure 7.19). By taking 12 copies of the $(2, 3, \infty)$ triangle, we can fill the fundamental region R for the free group F_2 generated by $z \mapsto 2 + z$ and $z \mapsto \frac{z}{2z+1}$ (Section 5.3). We know from Section 5.3 how to fill $\mathbb{H}^2$ with copies of R; hence, the same process also gives a tessellation by $(2, 3, \infty)$ triangles. Again, the tessellation is symmetric because it is generated by reflections in the sides of the triangle.

Exercises

7.3.1. Generalize the exercises in Section 7.1 to find a presentation of the orientation-preserving subgroup of the group generated by reflections in the sides of a (p, q, r) triangle.

7.3.2. Verify that the side identifications of a 14-gon described in Example 2 produce the result claimed.

7.3.3. The tessellation of $\mathbb{H}^2$ by $(2, 3, 7)$ triangles can be viewed as a subdivision of a tessellation by quadrilaterals like that shown in Figure 7.20.

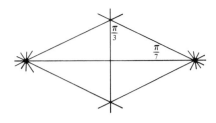

FIGURE 7.20.

Show that the latter tessellation is *not* symmetric by considering the angle conditions.

7.3.4. Without counting the triangles in Figure 7.18, show that 336 is the right number of $(2, 3, 7)$ triangles to make a hyperbolic surface of genus 3.

7.3.5. Show that a hyperbolic surface of genus 2 can be tessellated symmetrically by 96 $(2, 3, 8)$ triangles.

7.3.6. Show that reflections in the sides of the $(2, 3, \infty)$ triangle with vertices i, ω, ∞ induce the side-pairing transformations $z \mapsto 1 + z$ and $z \mapsto -1/z$ in its double. Verify that the Γ generated by these transformations has F_2 as a subgroup. What is the index of F_2 in Γ?

7.3.7. Show that the group Γ of Exercise 7.3.6 has a presentation

$$\langle g, h \mid g^2 = h^3 = 1 \rangle.$$

7.4 Poincaré's Theorem for Compact Polygons

Let g_i, for $i = 1, \ldots, n$, be the side-pairing transformations of a compact polygon Π which satisfies the side and angle conditions. That is, Π being given as a particular polygon in $\tilde{S}$ (equal to $\mathbb{S}^2$, $\mathbb{R}^2$, or $\mathbb{H}^2$), g_i is the orientation-preserving isometry that realizes the side pairing $s_i \to s_i'$ of Π. Ultimately, we shall show that if Γ is the group generated by $g_1, \ldots, g_n$, then Π is a fundamental region for the action of Γ on $\tilde{S}$. The problem is to show that $\tilde{S}$ is correctly tessellated by copies of Π. Instead, we shall construct a "correctly tessellated" surface $\tilde{S}_\Pi$ by brute force, associating tiles with elements of Γ and joining them according to the structure of Γ. When the tessellation of $\tilde{S}_\Pi$ is lifted to the $\mathbb{S}^2$, $\mathbb{R}^2$, or $\mathbb{H}^2$ known to cover $\tilde{S}_\Pi$ by the Killing–Hopf theorem, we find that the tile $(g)\Pi$ formally associated with $g \in \Gamma$ is indeed $g\Pi$; hence, Π is a fundamental polygon for Γ.

Construction of $\tilde{S}_\Pi$. *For each $g \in \Gamma$ take a copy $(g)\Pi$ of Π and an isometry $(g) : \Pi \to (g)\Pi$. We label the sides of $(g)\Pi$ by the same labels s_i,*

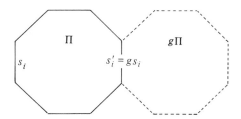

FIGURE 7.21.

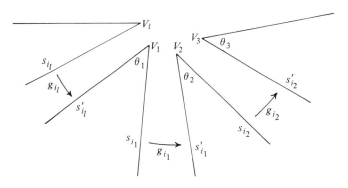

FIGURE 7.22.

s_i' as their preimages in Π. $\tilde{S}_\Pi$ is the identification space formed from these polygons by identifying the side s_i' of $(g)\Pi$ with the side s_i of $(gg_i)\Pi$.

This ensures, in particular, that $(gg_i)\Pi$ is adjacent to $(g)\Pi$ in the same way that $g_i\Pi$ is adjacent to Π in $\tilde{S}$ (Figure 7.21), for each $g \in \Gamma$.

Proposition. $\tilde{S}_\Pi$ is a complete geometric surface.

Proof. Certainly each interior point of each $(g)\Pi$ has a disc neighborhood. Since the sides of the polygons are identified in pairs, interior points of sides also have disc neighborhoods. The main problem is to show that vertices have disc neighborhoods.

Consider a vertex cycle $V_1, \ldots, V_l$ of Π (cf. Section 7.2). This means that the corners at $V_1, \ldots, V_l$ are related by a cyclic sequence $g_{i_1}, \ldots, g_{i_l}$ of side pairings as shown in Figure 7.22. By the angle condition which Π is assumed to satisfy, the angle sum $\theta_1 + \ldots + \theta_l$ of these corners is $2\pi/p$ for some positive integer p.

Now if $r_{\theta_j}(V_j)$ denotes the rotation about V_j through θ_j (in the sense from $s_{i_{j-1}}'$ to s_{i_j}), we have

$$g_{i_l} r_{\theta_l}(V_l) \cdots g_{i_2} r_{\theta_2}(V_2) g_{i_1} r_{\theta_1}(V_1) = 1 \tag{1}$$

because the isometry on the left-hand side is the identity on s'_{i_l} (remember to read the product of isometries from right to left) and is orientation-preserving (since the g_i are, by hypothesis). Also

$$r_{\theta_2}(V_2)g_{i_1} = g_{i_1}r_{\theta_2}(V_1)$$

because g_{i_1} sends V_1 to V_2 and therefore

$$r_{\theta_2}(V_2) = g_{i_1}r_{\theta_2}(V_1)g_{i_1}^{-1}.$$

Similarly,

$$r_{\theta_3}(V_3)g_{i_2}g_{i_1} = g_{i_2}g_{i_1}r_{\theta_3}(V_1),$$

and so on, eventually giving

$$g_{i_l}\ldots g_{i_1}r_{\theta_l}(V_1)\ldots r_{\theta_1}(V_1) = 1 \tag{2}$$

after all rotations have been brought to the right. But $r_{\theta_l}(V_1)\ldots r_{\theta_1}(V_1) = r_{\theta_l+\cdots+\theta_1}(V_1)$; hence, (2) says $g_{i_l}\ldots g_{i_1}$ is a rotation through $2\pi/p$.

It follows that $(g_{i_l}\ldots g_{i_1})^p$ is the least power of $g_{i_l}\ldots g_{i_1}$, which equals 1. Hence, the sequence of corners at a vertex of $\tilde{S}_\Pi$ corresponding to the cycle $V_1,\ldots,V_l$ (i.e., the corners corresponding to the sequence of group elements $g_{i_1}, g_{i_2}g_{i_1},\ldots, g_{i_l}\cdots g_{i_1}, g_{i_1}g_{i_l}\cdots g_{i_1}, g_{i_2}g_{i_1}g_{i_l}\cdots g_{i_1},\ldots$) closes up after precisely p repetitions of the cycle, i.e., when the angle sum is 2π, as required for a disc neighborhood.

This completes the proof that $\tilde{S}_\Pi$ is a geometric surface. The completeness of $\tilde{S}_\Pi$ follows by the argument of Section 5.7. □

Theorem (Poincaré [1882]). *A compact polygon Π satisfying the side and angle conditions is a fundamental region for the group Γ generated by the side-pairing transformations of Π.*

Proof. Construct the surface $\tilde{S}_\Pi$ as above. Since $\tilde{S}_\Pi$ is a complete geometric surface, by the proposition, it follows from the Killing–Hopf theorem and Section 6.5 that $\tilde{S}_\Pi = \tilde{S}/\pi_1(\tilde{S}_\Pi)$, where $\tilde{S}$ is $\mathbb{S}^2$, $\mathbb{R}^2$, or $\mathbb{H}^2$.

Assume first that $\pi_1(\tilde{S}_\Pi) = \{1\}$ (we shall later show that this is always the case). Then $\tilde{S}_\Pi = \tilde{S}$, and the tessellation of $\tilde{S}_\Pi$ by the polygons $(g)\Pi$ is a tessellation of the surface $\tilde{S}$ ($\mathbb{S}^2$, $\mathbb{R}^2$, or $\mathbb{H}^2$) containing the original Π, so we can take $(1)\Pi$ to be Π itself.

I claim that $(g)\Pi = g\Pi$ for any $g \in \Gamma$. This is proved by induction on k, where

$$g = g_{i_1}^{\epsilon_1}\ldots g_{i_k}^{\epsilon_k} \quad \text{and each } \epsilon_j = \pm 1.$$

The claim is true for $k = 1$ because the neighbors $g_i^{\pm 1}\Pi$ of Π are the $(g_i^{\pm 1})\Pi$ by definition of $\tilde{S}_\Pi$. Now assume that $(g')\Pi = g'\Pi$ when g' is a product of $\leq k - 1$ generators or their inverses. By definition of $\tilde{S}_\Pi$, $(g'g_i^{\pm 1})\Pi$ is to $(g')\Pi = g'\Pi$ what $g_i^{\pm 1}\Pi$ is to Π. That is,

$$(g'g_i^{\pm 1})\Pi = g'(g_i^{\pm 1}\Pi) = g'g_i^{\pm 1}\Pi,$$

and the induction is complete.

Thus, if $\pi_1(\tilde{S}_\Pi) = \{1\}$, the polygons $(g)\Pi$ tessellating $\tilde{S}_\Pi = \tilde{S} = \mathbb{S}^2$, $\mathbb{R}^2$, or $\mathbb{H}^2$ are precisely the polygons $g\Pi$ for $g \in \Gamma$; so Π is a fundamental region for Γ.

Now we stop assuming $\pi_1(\tilde{S}_\Pi) = \{1\}$ and consider the tessellation of $\tilde{S}$ obtained by lifting the tessellation of $\tilde{S}_\Pi = \tilde{S}/\pi_1(\tilde{S}_\Pi)$. We wish to show that only one polygon of $\tilde{S}$ lies over each polygon $(g)\Pi$ of $\tilde{S}_\Pi$, thus *proving* that $\pi_1(\tilde{S}_\Pi) = \{1\}$. Naturally we expect the one and only polygon of $\tilde{S}$ over $(g)\Pi$ to be $g\Pi$. Well, if $g = g_{i_1}^{\epsilon_1} \ldots g_{i_k}^{\epsilon_k}$, then the definition of $\tilde{S}_\Pi$ says we reach $(g)\Pi$ from $(1)\Pi$ by the sequence of edge crossings corresponding to $g_{i_k}^{\epsilon_k}, \ldots, g_{i_1}^{\epsilon_1}$. We reach $g\Pi$ from Π in $\tilde{S}$ as above, by exactly the same sequence of crossings; hence, any path from $(1)\Pi$ to $(g)\Pi$ in $\tilde{S}_\Pi$ lifts to a path from Π to $g\Pi$ in $\tilde{S}$. □

Remark. The above proof clearly identifies the role of the compactness of Π. It is needed to guarantee the completeness of the surface $\tilde{S}_\Pi$, and, hence, permit application of the Killing–Hopf theorem. The examples of the noncompact polygons Π in Exercises 7.2.1 and 7.2.2, for which $\tilde{S}_\Pi$ is *not* complete, show that the analogous theorem for noncompact polygons is false. The only way to get a "Poincaré's Theorem" for noncompact Π is to *assume* completeness of $\tilde{S}_\Pi$ or, what comes to the same thing, completeness of the surface S_Π which results from identifying the paired sides of Π. Indeed, Poincaré's theorem is often stated in the latter form, though it seems unfortunate to assume such a large part of what one wants to prove.

Exercises

7.4.1 (Poincaré [1882]). By considering a Cayley graph for the group Γ generated by the side-pairing transformations of Π, or otherwise, show that Γ is defined by the relations

$$(g_{i_l}^{\epsilon_l} \ldots g_{i_1}^{\epsilon_1})^p = 1$$

mentioned in the proof of the proposition.

7.4.2. Suppose that S_Π is a surface on which the vertices with angle sums $\neq 2\pi$ are $V(p_1), \ldots, V(p_m)$ with respective angle sums $2\pi/p_1, \ldots, 2\pi/p_m$ for some integers $p_1, \ldots, p_m > 1$. Using the normal form for orientable surfaces, show that S_Π is also the identification space S_N of a polygon N with boundary $c_1 c_1^{-1} \ldots c_m c_m^{-1} a_1 b_1 a_1^{-1} b_1^{-1} \ldots a_n b_n a_n^{-1} b_n^{-1}$. (Figure 7.23).

7.4.3 (Klein [1882]). Deduce that the group Γ generated by the side-pairing transformations of Π has a presentation

$$\langle c_1, \ldots, c_m, a_1, \ldots, a_n, b_1, \ldots, b_n \mid c_1^{p_1} = \cdots = c_m^{p_m}$$
$$= a_1 b_1 a_1^{-1} b_1^{-1} \ldots a_n b_n a_n^{-1} b_n^{-1} = 1 \rangle.$$

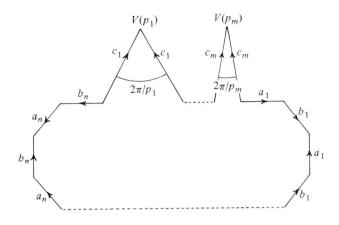

FIGURE 7.23.

7.5   Discussion

Poincaré [1880] first proved his theorem for a certain convex polygon in $\mathbb{D}^2$ which arose from a differential equation. It appears that he rediscovered Beltrami's route from $\mathbb{D}^2$ to $\mathbb{P}^2$ (Figures 4.24 and 4.25) and saw that it implies the convexity of polygons whose angles are less than π. Using this observation, he was able to show that $\mathbb{D}^2$ could be tessellated by copies of his polygon without overlapping. In his [1882] he claimed the theorem for general polygons satisfying the side and angle conditions, but his proof is not now regarded as satisfactory. Poincaré did not have an adequate grasp of the universal covering, which was later seen to be the key to a rigorous proof.

The analysis of spherical and euclidean tessellations implicit in Exercises 7.1.1 to 7.1.6 can be extended to find all the (finitely many) possible symmetry groups of the sphere and euclidean plane. See, for example, Lyndon [1985] or Montesinos [1987]. The complete enumeration of these symmetries in the 19th century was an important factor in the recognition of group theory as the "theory of symmetry". The exercises also illustrate the intimate relation between tessellations and combinatorial group theory—the theory of generators and relations. This theory began in the work of Klein [1882] and his student Dyck [1882]. A wealth of material on the combinatorial group theory of tessellations is collected in the book of Magnus [1974].

The amazing $(2, 3, 7)$ tessellation and the surface that Klein built from 336 of its triangles are illuminated by the article of Gray [1982]. They are also discussed at greater length in his book [1986] and, from a somewhat different viewpoint, in Jones and Singerman [1987].

Groups generated by reflections have a certain advantage over orientation-preserving groups, inasmuch as they have natural fundamental regions. One

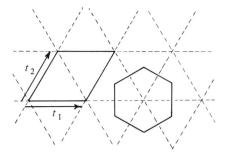

FIGURE 7.24.

can always find a region bounded by lines of reflection, as we have for triangle groups. In contrast, the fundamental regions for a group of translations are somewhat arbitrary. For example, both the regions in Figure 7.24 are fundamental regions for the euclidean group generated by $t_1(z) = z + 1$, $t_2(z) = z + \frac{1+\sqrt{3}}{2}$.

8
Tessellations of Compact Surfaces

8.1 Orbifolds and Desingularizations

Poincaré's theorem for a compact polygon Π satisfying the side and angle conditions tells us that the identification space S_Π, obtained by identifying sides of Π according to the side pairing, is also an orbit space $\tilde{S}/\Gamma$. Here $\tilde{S} = \tilde{S}_\Pi$ is $\mathbb{S}^2$, $\mathbb{R}^2$, or $\mathbb{H}^2$—the surface from which Π originates—and Γ is the group generated by the side-pairing transformations of Π. Because of its interpretation as an orbit space, S_Π is also called an *orbifold*.

An orbifold S_Π is topologically a two-dimensional manifold, but it is not a geometric surface unless the angle sum for each cycle of identified vertices is 2π (cf. Section 5.5). Nevertheless, its neighborhood structure is certainly close to that of a geometric surface, being different only at the finitely many points corresponding to vertex cycles whose angle sums are less than 2π. The neighborhood of such a point, instead of being isometric to a disc, is isometric to a *cone*. We call such a point a *cone point* or a *cone singularity*.

A *cone of angle* $2\pi/n$, for an integer $n > 1$, is the identification space of a disc sector of angle $2\pi/n$, the identification map $a \to a'$ being rotation through $2\pi/n$ (Figure 8.1). Equivalently, of course, the cone can be defined as the orbit space $D/\mathbb{Z}_n$, where D is the disc and $\mathbb{Z}_n$ is the group generated by a rotation of $2\pi/n$ about the center of D. D can be a disc of $\mathbb{S}^2$, $\mathbb{R}^2$, or $\mathbb{H}^2$, the ordinary cone being the case $D \subset \mathbb{R}^2$. The orbit map $\mathbb{Z}_n \cdot : D \to D/\mathbb{Z}_n$ is called a *desingularization* of the cone because it "lifts" the cone to a nonsingular (i.e., geometric) surface, the disc D (Figure 8.2).

In general, we shall define a *desingularization of an orbifold* $S_\Pi = \tilde{S}_\Pi/\Gamma$ (where $\tilde{S}_\Pi = \mathbb{S}^2$, $\mathbb{R}^2$, or $\mathbb{H}^2$ as in Section 7.4) to be a map $\Gamma/\Gamma' \cdot : \tilde{S}_\Pi/\Gamma' \to \tilde{S}_\Pi/\Gamma$, where $\tilde{S}_\Pi/\Gamma'$ is a geometric surface, Γ' is a subgroup of Γ, and $\Gamma/\Gamma' \cdot$ is defined as follows. Let

$$\Gamma = \Gamma' g_1 \cup \Gamma' g_2 \cup \ldots$$

be the partition of Γ into distinct Γ'-cosets. This gives a partition of the Γ-orbit $\Gamma P \in \tilde{S}_\Pi/\Gamma$ of a point $P \in \tilde{S}_\Pi$ into the Γ'-orbits $\Gamma' g_1 P$, $\Gamma' g_2 P, \ldots \in \tilde{S}_\Pi/\Gamma'$ of the points $g_1 P$, $g_2 P, \ldots \in \tilde{S}_\Pi$. $\Gamma/\Gamma' \cdot$ is the map that sends all $\Gamma' g_i P$ to ΓP. In the desingularization of the cone above we have $\Gamma = \mathbb{Z}_n$

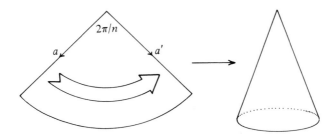

FIGURE 8.1.

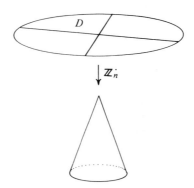

FIGURE 8.2.

and $\Gamma' = \{1\}$. (We call the map Γ/Γ'· because it sends $\Gamma'P$ to ΓP, hence it is formally "multiplication on the left by Γ/Γ'".)

One can regard $\{\Gamma'g_1P, \Gamma'g_2P, \ldots\}$ as the "coset orbit" of each of its points $\Gamma'g_iP \in \tilde{S}_\Pi/\Gamma'$, so with this generalization of the notion of orbit, Γ/Γ'· is an orbit map. It reduces to the ordinary notion of orbit map when the cosets $\Gamma'g_i$ form a group, i.e., when Γ' is a normal subgroup of Γ. Normality also turns out to be important for another reason. It is precisely the condition for the tessellation of $\tilde{S}_\Pi/\Gamma'$ by copies of Π ("projected" from $\tilde{S}_\Pi$ or "lifted" from $\tilde{S}_\Pi/\Gamma$) to be symmetric.

Theorem. *If $\tilde{S}_\Pi/\Gamma' \to \tilde{S}_\Pi/\Gamma$ is a desingularization, then the tessellation of $\tilde{S}_\Pi/\Gamma'$ by copies of Π is symmetric if and only if Γ' is a normal subgroup of Γ.*

Proof. To gain a clearer view of the surface $\tilde{S}_\Pi/\Gamma'$, we recall from Remark (2) of Section 7.2 that Γ' has a fundamental region of the form

$$\Pi' = g_1\Pi \cup g_2\Pi \cup \ldots$$

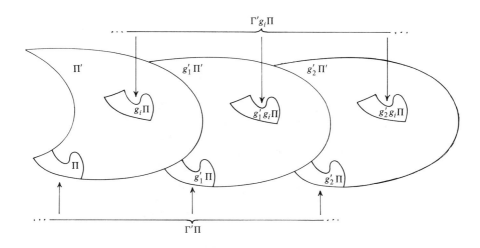

FIGURE 8.3.

where

$$\Gamma = \Gamma' g_1 \cup \Gamma' g_2 \cup \ldots$$

is the decomposition of Γ into Γ'-cosets. Thus, $\tilde{S}_\Pi$ is tessellated by the regions $g'\Pi'$ as g' runs through Γ'. $\tilde{S}_\Pi/\Gamma'$ is the result of identifying paired sides of Π'; hence, it is tiled by copies of Π corresponding to the cosets of Γ' in Γ.

A tile of $\tilde{S}_\Pi/\Gamma'$ is the Γ'-orbit

$$\Gamma' g_i \Pi = \{ g' g_i \Pi \mid g' \in \Gamma' \}$$

of a tile $g_i\Pi$ of Π'. The Γ'-orbits of Π and of an arbitrary $g_i\Pi$ are shown schematically in Figure 8.3. Thus, symmetry of the tessellation of $\tilde{S}_\Pi/\Gamma'$ means that the Γ'-orbit $\Gamma'\Pi$ can be mapped onto any other Γ'-orbit, $\Gamma' g\Pi$ for an arbitrary $g \in \Gamma$, by some $h \in \Gamma$. But if $h\Gamma'\Pi = \Gamma' g\Pi$, then $hg' = g$, where $g'\Pi$ is the tile that h sends to $g\Pi$. Since $g' \in \Gamma'$, it follows that $g'\Gamma' = \Gamma'$ and therefore

$$g\Gamma' = hg'\Gamma' = h\Gamma' = \Gamma' g,$$

so Γ' is a normal subgroup of Γ.

Conversely, if $g\Gamma' = \Gamma' g$ for any $g \in \Gamma$, then g maps the Γ'-orbit $\Gamma'\Pi$ onto the arbitrary Γ'-orbit $\Gamma' g\Pi$; so the tessellation of $\tilde{S}_\Pi/\Gamma'$ is symmetric. $\square$

Remarks. (1) We did not actually assume that $\tilde{S}_\Pi/\Gamma'$ was a geometric surface in the proof above; hence, the theorem holds for maps $\tilde{S}_\Pi/\Gamma' \to \tilde{S}_\Pi/\Gamma$ between arbitrary orbifolds. Such a map is called a *branched covering* because it is a local isometry, except possibly at vertices of $\tilde{S}_\Pi/\Gamma$ where it may "branch," i.e., map a point of $\tilde{S}_\Pi/\Gamma'$ onto a cone point of $\tilde{S}_\Pi/\Gamma$ with smaller angle.

(2) If $\tilde{S}_\Pi/\Gamma'$ *is a geometric surface, then* Γ' must act on $\tilde{S}_\Pi$ without fixed points. Since the elements of Γ' are products of side-pairing transformations of Π, and hence orientation-preserving, it follows that they must be translations or limit rotations. Γ' is therefore *torsion-free*, i.e., if $1 \neq g' \in \Gamma'$, then $g'^n \neq 1$ if $n \neq 0$. Conversely, suppose Γ' is a torsion-free subgroup of Γ. Then Γ' contains no rotations, and since its elements belong to Γ they are orientation-preserving and hence without fixed points. Γ acts discontinuously on $\tilde{S}$, since there is a well-defined distance between orbits, hence so does Γ'. Thus $\tilde{S}_\Pi/\Gamma'$ is a geometric surface.

(3) $\tilde{S}_\Pi/\Gamma'$ is compact if and only if the fundamental region Π' for Γ' consists of finitely many copies $g_i\Pi$ of Π. In other words, there are finitely many Γ'-cosets in Γ, i.e., Γ' is of *finite index* in Γ. [One may also recall, from Remark (2) of Section 7.2, that Π' may be taken to be a polygon in this case.]

Now the problem we wish to solve in this chapter is the following: *given a compact polygon Π satisfying the side and angle conditions* (7.2), *find a compact orientable surface symmetrically tessellated by copies of* Π. In view of the theorem, and Remarks (2) and (3), this problem may be restated purely algebraically: *find a normal, torsion-free subgroup of finite index in the group Γ generated by the side-pairing transformations of* Π.

Restating the problem algebraically helps to the following extent. One can proceed, by elementary group theory, from an arbitrary finite-index subgroup Γ' of Γ to a subgroup Γ^* of Γ' which is normal in Γ and still of finite index. Consequently, if $\tilde{S}_{\Pi'}/\Gamma'$ is a compact surface, then $\tilde{S}_\Pi/\Gamma^*$ is a compact surface symmetrically tessellated by copies of Π. Thus, the problem is reduced to the slightly easier one of finding any torsion-free subgroup Γ' of Γ of finite index, i.e., of finding any desingularization of $\tilde{S}_\Pi/\Gamma$ by a compact surface. The latter problem is still difficult, so we shall postpone it until we have carried out the group-theoretic reduction.

Exercises

8.1.1. Illustrate the theorem above in the case where Π is the $(2, 4, 4)$ triangle (so $\tilde{S}_\Pi = \mathbb{R}^2$) and

$\Gamma = \{\text{orientation-preserving symmetries of the square tessellation}\}$,

$\Gamma' = \{\text{translations in } \Gamma\}$.

8.1.2. Let $\Gamma = \langle a_1, b_1, \ldots, a_n, b_n \mid a_1 b_1 a_1^{-1} b_1^{-1} \ldots a_n b_n a_n^{-1} b_n^{-1} = 1 \rangle$ (cf. Section 6.6). Show that any group G on generators $a_1, \ldots, a_n$ is obtainable by adding to Γ the relations $b_1 = \ldots = b_n = 1$ plus relations of the form $w_i(a_1, \ldots, a_n) = 1$. Equivalently, $G = \Gamma/\Gamma'$, where Γ' is the kernel of the homomorphism defined by $b_1, \ldots, b_n \mapsto 1$, $w_i(a_1, \ldots, a_n) \mapsto 1$.

8.1.3. Deduce from Exercise 8.1.2 that any finitely generated group is the symmetry group of a tessellated surface, and that any finite group is the symmetry group of a tessellated compact surface.

8.2 From Desingularization to Symmetric Tessellation

The group-theoretic steps from desingularisation to symmetry are the following:

(1) If Γ' is a subgroup of Γ for which $\tilde{S}_\Pi/\Gamma'$ is a compact geometric surface, then the fundamental region Π' for Γ' consists of a finite number, N, of copies of Π. (Since Γ is orientation-preserving, the surface $\tilde{S}_\Pi/\Gamma'$ is also orientable.)

(2) There are only finitely many subgroups of Γ whose fundamental region consists of N copies of Π [e.g., because the regions may be taken to be adjacent by Remark (2) of Section 7.2]. These include the conjugates $g^{-1}\Gamma'g$ because conjugates all have the same index (conjugation by g is an isomorphism). Hence, there are only finitely many conjugates of Γ'.

(3) If Γ', Γ'' are subgroups of finite index in Γ, then $\Gamma'\cap\Gamma''$ is also of finite index (for a proof, see Poincaré's lemma below). Hence, by (2), the intersection Γ^* of the conjugates of Γ' is of finite index. Also, because $\Gamma^* \subset \Gamma'$ and $\tilde{S}_\Pi/\Gamma'$ is a compact, orientable geometric surface, the elements of Γ^* are translations by Sections 2.9 and 6.9.

(4) Since Γ^* consists of translations, $\tilde{S}_\Pi/\Gamma^*$ is a geometric surface, and, hence, the coset orbit map $\tilde{S}_\Pi/\Gamma^* \to \tilde{S}_\Pi/\Gamma$ is a desingularization. Since Γ^* is of finite index, by (3), $\tilde{S}_\Pi/\Gamma^*$ is compact by Remark (3) of Section 8.1. Finally, since Γ^* is closed under conjugation, by (3), it is a normal subgroup of Γ and, hence, the tessellation of $\tilde{S}_\Pi/\Gamma^*$ is symmetric by the theorem in Section 8.1.

Poincaré's Lemma (Poincaré [1887]). *If Γ' and Γ'' are subgroups of finite index in Γ, then so is $\Gamma'\cap\Gamma''$.*

Proof. Consider a coset $(\Gamma'\cap\Gamma'')g$ of the subgroup $\Gamma'\cap\Gamma''$. Obviously, $(\Gamma'\cap\Gamma'')g \subseteq \Gamma'g\cap\Gamma''g$, but also $\Gamma'g\cap\Gamma''g \subseteq (\Gamma'\cap\Gamma'')g$ because if $g'g \in \Gamma'g$ and $g'g = g''g \in \Gamma''g$, then $g' = g'' \in \Gamma'\cap\Gamma''$. Hence, each coset of $\Gamma'\cap\Gamma''$ is of the form $\Gamma'g\cap\Gamma''g$.

Now if Γ' is of index m, say, then $\Gamma'g$ takes m values as g runs through Γ. Similarly, $\Gamma''g$ takes, say, n values. Therefore, $\Gamma'g \cap \Gamma''g$ takes at most mn values, so $\Gamma' \cap \Gamma''$ has at most mn cosets. $\square$

Exercise

8.2.1. Give subgroups Γ', Γ'' of $\Gamma = \pi_1$ (torus) for which

$$\text{index of } \Gamma' \cap \Gamma'' = (\text{index of } \Gamma')(\text{index of } \Gamma'').$$

(It is obvious that inequality can hold. Why?)

8.3 Desingularizations as (Branched) Coverings

Now we return to the main problem: finding a desingularization of an orbifold $S_\Pi = \tilde{S}_\Pi/\Gamma$. The view of S_Π as the orbit space of Γ is *not* very helpful here. It is more profitable to view S_Π as the result of identifying paired sides of Π and to view a desingularization as a literal "covering" of S_Π by suitably joined copies of Π. This view, which comes from the classical theory of Riemann surfaces, can be introduced with the example of the cone $D/\mathbb{Z}_n$.

One imagines the cone covered by n copies of itself, called *sheets*. Each sheet is slit along a line from the apex of the cone, and the left side of the cut in the ith sheet is joined to the right side of the cut in the $(i+1)$th sheet (where $i + 1$ is taken mod n). This is intrinsically the same as cyclically joining n fundamental sectors for $\mathbb{Z}_n$ to form D (Figure 8.4); however, it has the advantage of showing the preimages

$$\tilde{P}_1, \ldots, \tilde{P}_n \text{ of a point } P = \{\tilde{P}_1, \ldots, \tilde{P}_n\} \in D/\mathbb{Z}_n \text{ really lying over } P.$$

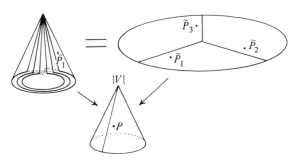

FIGURE 8.4.

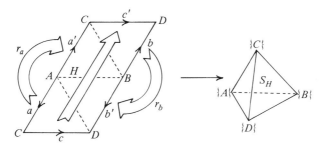

FIGURE 8.5.

More importantly, it shows how a cone point $\{V\}$ of angle $2\pi/n$ can be desingularized by "unwrapping" n sheets joined cyclically over $\{V\}$. The "unwrapped" sheets form the disc which is mapped onto a neighborhood of $\{V\}$ by the $\mathbb{Z}_n$-orbit map. (Since the points of the cone are $\mathbb{Z}_n$-orbits, i.e., *sets* of points of D, it is correct to call the apex of the cone $\{V\}$ rather than V.)

This view of desingularization as "unwrapping" becomes invaluable when we have an orbifold S_Π with more than one cone point to desingularize. We can form a multiple cover of S_Π, then cut and rejoin the sheets so as to "unwrap" all cone points simultaneously. As mentioned in Remark (1) of Section 8.1, this kind of covering is said to be *branched* over the cone points it unwraps.

As an example, consider the tetrahedron, which is a euclidean orbifold S_H obtained by pairing sides of a hexagon H (four of the sides are really "half-sides" of a parallelogram) as shown in Figure 8.5. The cone points $\{A\}$, $\{B\}$, $\{C\}$, $\{D\}$ of S_H (where $\{X\}$ denotes the set of all points labeled X on H) all have angle π; hence, they are desingularized by a 2-sheeted covering. The cyclic joining of the sheets can be achieved at each vertex of S_H simultaneously by cutting the tetrahedral sheets along $\{A\}\{D\}$ and $\{B\}\{C\}$ and joining the left side of the cut on one sheet to the right side of the corresponding cut on the other sheet. If we distinguish lines on the two sheets by subscripts 1 and 2, as in Figure 8.6, this means joining a_1 to a_2', a_2 to a_1', b_1 to b_2' b_2 to b_1'. (The pairing of an unprimed letter with its primed form reflects the side pairing of the original polygon. The different subscripts reflect the different sheets being joined.)

The desingularizing surface can be seen to be a torus by separating the cut sheets, which are topologically cylinders, before making the joins (Figure 8.7).

To see that this is a desingularization in the sense of Section 8.1, let Γ be the group generated by the side-pairing transformations r_a, r_b, t_c of H, so $S_H = \tilde{S}_H/\Gamma$, where in this case $\tilde{S}_H = \mathbb{R}^2$. We now use the covering to construct a fundamental region H' for a subgroup Γ' of Γ such that the

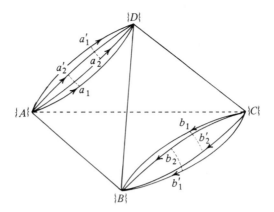

FIGURE 8.6.

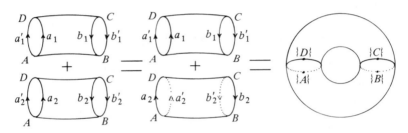

FIGURE 8.7.

coset orbit map $\mathbb{R}^2/\Gamma' \to \mathbb{R}^2/\Gamma$ is precisely the covering map.

The sheets of the covering surface are two copies H_1, H_2 of H, with edges joined as shown in Figure 8.6. We form H' by joining H_1, H_2 only so far as to form a polygon, the remaining joins being left as a side pairing of H'. For example, one can join b_1 to b_2' and b_2 to b_1' (Figure 8.8). The covering map is indicated by the preimages $\tilde{P}_1 \in H_1$ and $\tilde{P}_2 \in H_2$ of a typical point $P \in H$. We now show that $\tilde{P}_1, \tilde{P}_2 \mapsto P$ is precisely the coset orbit map $\mathbb{R}^2/\Gamma' \to \mathbb{R}^2/\Gamma$ for the group Γ' generated by the side pairings of the polygon H'.

The side pairings of H' are the translations t_c (which maps c_1 onto c_1' and c_2' onto c_2) and $r_b r_a$ (which maps a_1 onto a_2 and a_1' onto a_2). Hence, the group Γ' they generate is certainly a subgroup of Γ. Moreover, because Γ' also contains $(r_b r_a)^{-1} = r_a^{-1} r_b^{-1} = r_a r_b$, and because $r_a r_b \cdot r_b = r_a$, it follows that $\Gamma' r_b$ contains r_a and r_b. Hence, we have the coset decomposition

$$\Gamma = \Gamma' \cap \Gamma' r_b,$$

and because $\tilde{P}_2 = r_b \tilde{P}_1$, the map $\tilde{P}_1, \tilde{P}_2 \mapsto P$ is indeed the coset orbit map $\mathbb{R}^2/\Gamma' \to \mathbb{R}^2/\Gamma$.

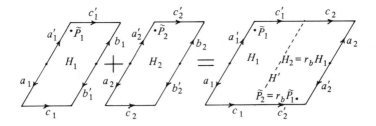

FIGURE 8.8.

Remark. The description of a branched covering of an orbifold S_Π as a pattern of cut and rejoined sheets can always be translated into the group-theoretic form $\tilde{S}_\Pi/\Gamma' \to \tilde{S}_\Pi/\Gamma$ via a fundamental polygon Π' for Γ'. Π' is a union of adjacent copies of Π in $\tilde{S}_\Pi$, each copy being a sheet and each adjacency being a join. Each side of Π' is paired with another found by following paired sides through the copies of Π in Π' until one reaches the boundary of Π'. In particular, each side pairing of Π' is a product of side pairings of Π and, hence, the group Γ' is a subgroup of Γ. Finally, the different copies of Π in Π' represent the different cosets of Γ' in Γ [by Remark (2) of Section 7.2, which also tells us that Π' *is* a polygon]; hence, the covering map is precisely the coset orbit map.

We shall take advantage of this fact in future by describing desingularizations in the more easily visualized form of branched coverings, omitting the construction of Γ'. We can also take the liberty of cutting sheets along curves not necessarily corresponding to the original edges of Π. This is permissible because any way of cutting S_Π down to a polygon merely yields another fundamental region for the same group Γ.

Exercises

8.3.1. Use $(6,6,6)$ triangles to construct a hyperbolic tetrahedron S_Π with cone angles $\pi/2$. Hence, show that S_Π is desingularized by a surface of genus 3. (Hint: Compute the Euler characteristic of a suitable cover.)

8.3.2. Consider the *modular orbifold* $\mathbb{H}^2/\Gamma$, where Γ is the group generated by the side-pairing transformations $z \mapsto 1 + z$ and $z \mapsto -1/z$ of the quadrilateral shown in Figure 7.19. Show that $\mathbb{H}^2/\Gamma$ has a cone point A of angle π and a cone point B of angle $2\pi/3$.

8.3.3. Desingularize the modular orbifold by a 6-fold branched covering, making disjoint cuts from A and B to ∞ and joining the sheets in three pairs over A and two triples over B. Compare the region Π' for the covering with the fundamental region for F_2 (cf. Section 7.3).

8.3.4. Generalize the idea of Exercise 8.3.3 to show that any orbifold S_Π

with noncompact Π and cone points of angles $2\pi/p_1, \ldots, 2\pi/p_n$ may be desingularized by a branched covering of $\operatorname{lcm}(p_1, \ldots, p_n)$ sheets.

8.4   Some Methods of Desingularization

As has already been suggested, the desingularization of an arbitrary orbifold S_Π is a complicated process. It is not, in general, feasible to find a branched covering of S_Π by a geometric surface in one stroke. Instead, one constructs a series of branched coverings of S_Π by other orbifolds, progressively simplifying or eliminating cone points. The following four methods enable us to simplify cone points until only one special type of desingularization (described in the next section) remains to be achieved.

To shorten the description of cone points we say a cone point has *index* p if the cone angle is $2\pi/p$. It also goes without saying that each cut between cone points must be a simple arc and disjoint from any other cuts.

8.4.1   Method 1. Elimination of Two Cone Points of the Same Index

If each cone point P, Q has index n, cover by n sheets, cut along an arc PQ from P to Q, and join the sheets cyclically along PQ (i.e., left of sheet 1 to right of sheet 2, left of sheet 2 to right of sheet 3, $\ldots$, left of sheet n to right of sheet 1, or $1 \to 2 \to \ldots \to n \to 1$ for short).

Generalization: Simultaneous Elimination. For any integer $m \geq 1$, we can create m copies of the original covering by taking mn sheets and joining them in m n-cycles along PQ [i.e., join $1 \to 2 \to \ldots \to n \to 1$, $n+1 \to n+2 \to \ldots \to 2n \to n+1, \ldots, m(n-1)+1 \to m(n-1)+2 \to \ldots \to mn \to m(n-1)+1$]. If P_1, Q_1 have index n_1 and P_2, Q_2 have index n_2, this idea enables us to simultaneously eliminate P_1, Q_1, P_2, Q_2 as follows. Cover with $n = \operatorname{lcm}(n_1, n_2)$ sheets, cut along disjoint arcs P_1Q_1 and P_2Q_2, then join the sheets in n/n_1 n_1-cycles over P_1Q_1 and in n/n_2 n_2-cycles along P_2Q_2; similarly for any number of pairs, using the lcm of their indices.

8.4.2   Method 2. Elimination of Three Cone Points of the Same Odd Index

If the cone points P, Q, R each have index $2n + 1$, cover by $2n + 1$ sheets and make cuts between P, Q, R as shown in Figure 8.9. Note that if we join the $2n + 1$ sheets cyclically in the same way along PR and QR this desingularizes not only the cone points P and Q, but also the cone point R. Joining the left of the ith sheet to the right of the $(i + 1)$th along PQ,

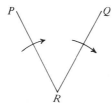

FIGURE 8.9.

then the left of the $(i+1)$th to the right of the $(i+2)$th along QR, means that in making a circuit around R one passes from the ith sheet to the $(i+2)$th (mod $2n+1$ of course). Hence, the sheets are joined over R in the $(2n+1)$-cycle $1 \to 3 \to 5 \to \ldots \to 2n+1 \to 2 \to 4 \to \ldots \to 2n \to 1$.

Generalization: Simultaneous Elimination. As with Method 1 we can create m copies of the original covering for any integer $m \geq 1$, and simultaneously eliminate several triples, or pairs and triples, of different indices by suitable choice of m. For example, to eliminate a pair P_1, Q_1 of index n_1 and a triple P_2, Q_2, R_2 of odd index n_2, we cover with $n = \mathrm{lcm}(n_1, n_2)$ sheets, then join the sheets in n/n_1 n_1-cycles over the cut P_1Q_1 and in n/n_2 n_2-cycles over the cuts P_2R_2 and Q_2R_2.

8.4.3   Method 3. Reducing the Number of Cone Points of Even Index to at Most One

If there are two cone points, A and B, of even index, let $2d$ be the greatest common divisor of their indices. Cover with $2d$ sheets, cut from A to B, and cyclically join the sheets along the cut. This divides the indices of A, B by $2d$ and hence makes at least one of the resulting cone points A', B' of odd index. If there were cone points other than A, B originally, this process creates $2d$ copies of each of them; hence, they can be grouped in pairs and eliminated by Method 1. Using an even number of sheets, $2m$, for this elimination leaves us with just $2m$ copies of A' and $2m$ copies of B'. The latter cone points can then be grouped in pairs of the same index and eliminated by Method 1.

8.4.4   Method 4. Elimination of Cone Points on a Nonspherical Surface

If the surface is not homeomorphic to a sphere, then it contains a simple closed nonseparating curve c (e.g., the curve a_1 of the normal form of Section 6.1). By moving this curve slightly, if necessary, we can ensure that c does not pass through any cone points. We then cover by two sheets, cut them along c, and join one sheet to the other. Since c is nonseparating, the result is a connected surface, and with two copies of each of the original cone points. Thus, the cone points can now be eliminated by Method 1.

Exercises

8.4.1. For each genus $n \geq 1$ and index $p \geq 2$ find a hyperbolic polygon Π for which S_Π is a surface of genus n (topologically) with a single cone point, of index p.

8.4.2. Let S_Π be a torus with a single cone point, of index p, as found in Exercise 8.4.1. Use the desingularization methods above to cover S_Π by a geometric surface of genus p.

8.4.3. Construct topological spheres S_Π with

  (i) two cone points, of the same index $p \geq 2$, and

  (ii) three cone points, with arbitrary indices p, q, $r \geq 2$.

8.5   Reduction to a Permutation Problem

Thanks to Method 4 of Section 8.3, the only orbifolds which remain to be desingularized are topological spheres. A sphere cannot have a single cone point and it cannot have just two unless they have the same index. For if two cone points have different indices m, n, then a cut between them yields a 2-gon with different angles— $2\pi/m$ and $2\pi/n$—which does not exist in $\mathbb{S}^2$, $\mathbb{R}^2$, or $\mathbb{H}^2$. (Taking $m = 1$ in this argument, i.e., taking the first point to be an ordinary point, disposes of the possibility of a sphere with just one cone point.) If the sphere has two cone points of the same index they can be eliminated by Method 1.

Thus, the only orbifolds which remain to be desingularized are spheres with at least three cone points. It will now be shown that these orbifolds can also be desingularized, assuming the following hypothesis:

(p, q, r)-**Hypothesis.** *For any integers p, q, $r \geq 2$ there is a permutation σ_p of order p and a permutation σ_q of order q whose product $\sigma_q \sigma_p$ is of order r.*

The *order* of a permutation σ is the least power of σ which equals the identity. Apart from this, the only feature of σ needed in what follows is the *cycle decomposition*—the partition of the set (of sheets, in our case) on which σ acts into disjoint σ-orbits. A σ-orbit is permuted cyclically by σ; hence, the order of σ is the lowest common multiple (lcm) of the lengths of its cycles.

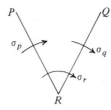

FIGURE 8.10.

Theorem. *Assuming the (p, q, r)-hypothesis, all orbifolds can be desingularized.*

Proof. By the remarks above, we can assume that the orbifold S_Π has at least three cone points P, Q, R. By Method 3 we can assume that at most one of the cone points has even index; let it be P if it exists at all.

We first show how all cone points can be eliminated when there are *more* cone points than P, Q, R. If the indices of P, Q, R are p, q, r, respectively, let σ_p, σ_q, σ_r be permutations (of s things, say) given by the (p, q, r)-hypothesis. We cover S_Π by s sheets, cut along PR and QR, and permute the sheets as indicated in Figure 8.10. [An arrow σ from left to right across one or more cuts indicates the joining of sheet i on the left to sheet $\sigma(i)$ on the right.]

Since σ_p has order p by hypothesis, it decomposes into cycles whose lengths $p_1, \ldots, p_l$ have lcm equal to p. In particular, $p_1, \ldots, p_l$ are divisors of p, so these cycles correspond to cone points $P_1, \ldots, P_l$ of respective indices $p/p_1, \ldots, p/p_l$ over P on the covering orbifold $S_{\Pi'}$. Similarly, there are points $Q_1, \ldots, Q_m$ of indices $q/q_1, \ldots, q/q_m$ over Q and points $R_1, \ldots, R_n$ of indices $r/r_1, \ldots, r/r_m$ over R.

As well as these cone points, on $S_{\Pi'}$ we have s copies of every other cone point originally on S_Π. We can therefore group the latter cone points on $S_{\Pi'}$ in pairs of equal index (if s is even) plus triples of equal index (if s is odd because $s \geq 2$ in any case) and, hence, eliminate them simultaneously by Methods 1 and 2 (because cone points other than P have odd index by hypothesis). By using an even number of sheets for this elimination (if necessary using the ability to multiply the sheet number in both Methods 1 and 2) we create multiple pairs of copies of the P_i, Q_j, R_k, which can likewise be eliminated by Method 1, thus desingularizing S_Π.

Now suppose, on the other hand, that P, Q, R are the only cone points on S_Π. Then the orbifold $S_{\Pi'}$, constructed as before, has just the P_i, Q_j, and R_k as its cone points. If these total more than three cone points, then we can desingularize $S_{\Pi'}$, and hence $S_{\Pi'}$ by the above argument. If not, then we have just one P_i, one Q_j and one R_k. That is, σ_p, σ_q, and σ_r are themselves cycles, so $S_{\Pi'}$ itself desingularizes S_Π. $\qquad \Box$

The (p, q, r)-hypothesis can also be stated in geometric terms. An equivalent hypothesis is that one can desingularize the orbifold $S(p, q, r)$ obtained by identifying paired sides of the double of a (p, q, r) triangle (see Exercise 8.5.1). In this sense, the general problem of desingularization reduces to the special problem for (p, q, r) triangles. However, as already mentioned in Section 7.3, the latter problem has no obvious geometric solution. It seems preferable to state the (p, q, r)-hypothesis in terms of permutations, as this paves the way for a group-theoretic solution.

As is well known, each element g of a finite group G can be interpreted as a permutation, namely, the permutation of G itself induced by multiplying each element of G by g (on the left, say). Thus, one way to construct the permutations σ_p, σ_q, σ_r required for the (p, q, r)-hypothesis would be to find a finite group G containing elements a, b such that a has order p, b has order q, and ba has order r. We shall do this in the next section.

Exercise

8.5.1. Show that desingularization of the orbifold $S(p, q, r)$ requires permutations σ_p, σ_q, σ_r with the properties stated in the (p, q, r)-hypothesis.

8.6   Solution of the Permutation Problem

It seems that the simplest finite groups containing two elements and their product with given orders are matrix groups over finite fields. Thus, to finish off the problem of desingularization, we have to administer another dose of algebra. This can be done very quickly if we assume the existence and basic properties of finite fields. In particular, we shall assume that there is a field $GF(s)$ of s elements, where s is any prime power, and that the multiplicative group of nonzero elements of $GF(s)$ is cyclic.

The group we consider is $\mathrm{PSL}(2, s)$, the group of substitutions of the form

$$f(z) = \frac{\alpha z + \beta}{\gamma z + \delta},$$

where α, β, γ, $\delta \in GF(s)$ and $\alpha\delta - \beta\gamma = 1$. As with linear fractional substitutions with real or complex coefficients, the behavior of f is largely reflected in its matrix

$$F = \begin{pmatrix} \alpha & \beta \\ \gamma & \delta \end{pmatrix}.$$

In fact, the key to the construction turns out to be the *trace* of F,

$$\mathrm{tr}(F) = \alpha + \delta,$$

and its invariance under conjugation. An easy calculation shows that

$$\mathrm{tr}(FG) = \mathrm{tr}(GF) \quad \text{for any matrices } F, G$$

and, therefore,

$$\text{tr}(GFG^{-1}) = \text{tr}(G^{-1}GF) = \text{tr}(F).$$

Following Feuer [1971] we base the construction of suitable group elements on the lemma below, which shows that the order of a certain type of substitution is determined by the trace of its matrix.

Lemma. *If s is an odd prime power and $\alpha + \delta = \rho + \rho^{-1}$ for some $\rho \neq 0, \pm 1$ of multiplicative order $2n$, then $f(z) = \frac{\alpha z + \beta}{\gamma z + \delta}$ has order n.*

Proof. The fixed points of f satisfy

$$\gamma z^2 + (\delta - \alpha)z - \beta = 0. \tag{1}$$

If $\gamma = 0$, then the determinant $1 = \alpha\delta - \beta\gamma$ is just $\alpha\delta$, so $\delta = \alpha^{-1}$ and $\alpha + \delta = \rho + \rho^{-1}$ implies $\alpha = \rho$, $\delta = \rho^{-1}$, or $\alpha = \rho^{-1}$, $\delta = \rho$. The matrix F of f is, therefore, $\begin{pmatrix} \rho & \beta \\ 0 & \rho^{-1} \end{pmatrix}$ or $\begin{pmatrix} \rho^{-1} & \beta \\ 0 & \rho \end{pmatrix}$. An easy induction shows

$$\begin{pmatrix} \rho & \beta \\ 0 & \rho^{-1} \end{pmatrix}^k = \begin{pmatrix} \rho^k & \frac{\rho^k - \rho^{-k}}{\rho - \rho^{-1}}\beta \\ 0 & \rho^{-k} \end{pmatrix};$$

hence, because $\rho - \rho^{-1} \neq 0$ and ρ^{2n} is the least power of ρ equal to 1 in $GF(s)$, by hypothesis, it follows that ρ^n is the least power equal to ρ^{-n}, and, hence, f has order n.

If $\gamma \neq 0$, then $2\gamma \neq 0$ because s is a power of an odd prime, and we can solve (1) by the ordinary quadratic formula

$$z = \frac{\alpha - \delta}{2\gamma} \pm \sqrt{\frac{(\alpha - \delta)^2 + 4\beta\gamma}{4\gamma^2}}.$$

These are two distinct roots z_1, z_2 because

$$\begin{aligned}
(\alpha - \delta)^2 + 4\beta\gamma &= (\alpha - \delta)^2 + 4\alpha\delta - 4 \quad \text{because } \alpha\delta - \beta\gamma = 1 \\
&= (\alpha + \delta)^2 - 4 \\
&= (\rho - \rho^{-1})^2 \quad \text{because } \alpha + \delta = \rho + \rho^{-1} \\
&\neq 0 \quad \text{because } \rho \neq 0, \pm 1.
\end{aligned}$$

Now consider the substitution $g(z) = \frac{z - z_1}{z - z_2}$ which sends $z_1 \mapsto 0$, $z_2 \mapsto \infty$. It follows that the substitution gfg^{-1} has fixed points 0, ∞ and hence is of the form $z' = kz$. But since

$$\text{tr}(GFG^{-1}) = \text{tr}(F) = \rho + \rho^{-1},$$

we must have $gfg^{-1}(z) = \rho z/\rho^{-1} = \rho^2 z$. This shows gfg^{-1} has order n, hence so has f. $\qquad\square$

Now we choose s to suit the orders p, q, r of the group elements we are trying to construct. Since the lemma gives us an element of order n for each $\rho \neq 0$, ± 1 in $GF(s)$ of order $2n$, we want elements ρ of orders $2p$, $2q$, $2r$. Since the multiplicative group of $GF(s)$ has $s - 1$ elements and is cyclic, suitable ρ will be present if $2pqr$ divides $s - 1$.

If $t > p, q, r$ is any prime (necessarily odd), then $\gcd(2pqr, t) = 1$, and the euclidean algorithm shows that t has an inverse under multiplication mod $2pqr$. Thus, t is in the finite group $\mathbb{Z}_{2pqr}^{\times}$ of integers under multiplication mod $2pqr$, and hence some power $t^m \equiv 1 \pmod{2pqr}$. That is, $2pqr$ divides $t^m - 1$. Then if we choose $s = t^m$, we shall have elements ρ of orders $2p$, $2q$, $2r$ in $GF(s)$.

At last we are ready for the theorem that completes the solution of the desingularization problem (Section 8.5) and hence, by Section 8.2, gives a symmetric tessellation of a compact surface by any compact polygon Π satisfying the side and angle conditions.

Theorem. *For any integers $p, q, r \geq 2$, there is an s such that $\mathrm{PSL}(2, s)$ contains elements a, b, ba of orders p, q, r, respectively.*

Proof. By the remarks above, we can find an odd prime power s such that $2pqr$ divides $s - 1$, so that $GF(s)$ contains elements α, β, γ of orders $2p$, $2q$, $2r$, respectively.

We consider the matrices

$$A = \begin{pmatrix} \alpha & \lambda \\ 0 & \alpha^{-1} \end{pmatrix}, \quad B = \begin{pmatrix} \beta + \nu & 1 \\ \mu & \beta^{-1} - \nu \end{pmatrix}.$$

Since $\mathrm{tr}(A) = \alpha + \alpha^{-1}$, $\mathrm{tr}(B) = \beta + \beta^{-1}$, it will suffice, by the lemma, to determine λ and ν so that $\mathrm{tr}(BA) = \gamma + \gamma^{-1}$. The extra parameter μ is there to enable us to satisfy the condition $\det(B) = 1$.

The condition $\det(B) = 1$ gives the equation

$$\nu^2 + \nu(\beta - \beta^{-1}) + \mu = 0. \tag{2}$$

We can find a nonzero μ satisfying (2) by choosing any $\nu \neq 0$, $\beta^{-1} - \beta$.

The condition $\mathrm{tr}(BA) = \gamma + \gamma^{-1}$ gives the equation

$$\alpha\beta + \alpha^{-1}\beta^{-1} + \nu(\alpha - \alpha^{-1}) + \lambda\mu = \gamma + \gamma^{-1}. \tag{3}$$

Having chosen ν, μ to satisfy (2) with $\mu \neq 0$, we get a unique solution of (3) for λ, so the elements a, b, ba of $\mathrm{PSL}(2, s)$ with matrices A, B, BA then have orders p, q, r by the lemma. $\square$

8.7 Discussion

The idea of branched coverings began with Riemann [1851]. To visualize the behavior of the multivalued "function" $\sqrt{z}$, Riemann imagined the plane $\mathbb{C}$ to be covered by two "sheets" with a branch point at 0. Lying above the point $z \in \mathbb{C}$ are the two values of $\sqrt{z}$ (Figure 8.11), one the negative of the other. This covering is topologically the same as the desingularization of a cone of angle π, though not, of course, geometrically, because the "cone" in this case is the plane $\mathbb{C}$ and the covering map (squaring) is nowhere a local isometry.

Like ordinary (unbranched) coverings (Section 6.10), branched coverings are also significant from the purely topological viewpoint. If, for example, we extend the square root covering of $\mathbb{C}$ to $\mathbb{C} \cup \{\infty\}$, we find that there is a similar branch point at ∞. We therefore have a 2-sheeted covering of the sphere $\mathbb{C} \cup \{\infty\}$, with two branch points. By unwrapping this covering surface, one easily finds it to be topologically a sphere. A more interesting example is the branched covering of $\mathbb{C} \cup \{\infty\}$ for the "function" $\sqrt{z(z-\alpha)(z-\beta)}$, or $1/\sqrt{z(z-\alpha)(z-\beta)}$. This covering is also 2-sheeted, because of the two values of the square root, but it has branch points at 0, α, β, and ∞. It is therefore topologically the same as the branched covering of the tetrahedron in Section 8.3, which we found to be a torus.

This finally explains how the elliptic function f of Section 2.10 maps onto a torus. We say that f maps the torus $\mathbb{C}/\Gamma$ onto the sphere $\mathbb{C} \cup \{\infty\}$, in general sending two orbits to the same complex number. It is therefore appropriate to view f as mapping to a 2-sheeted covering of the sphere. It happens that there are four branch points, so if the covering is peeled off the sphere, f does indeed map to a torus. The four branch points in fact belong to a "function" of the form $1/\sqrt{z(z-\alpha)(z-\beta)}$ whose integral is the elliptic integral inverse to f. (For more information on elliptic functions see Gray [1986] or Jones and Singerman [1987].)

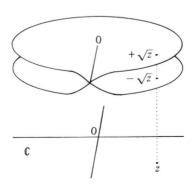

FIGURE 8.11.

The finite groups $\mathrm{PSL}(2, s)$ which intrude at the end of this story are not as foreign as they may seem at first. The smaller ones of the form $\mathrm{PSL}(2, p)$ [or $\mathrm{PSL}(2, \mathbb{Z}_p)$ as they are usually written] have a lot to do with classical tessellations. $\mathrm{PSL}(2, 5)$ is the icosahedral group A_5 and $\mathrm{PSL}(2, 7)$ is the symmetry group of Klein's surface with the 336 $(2, 3, 7)$ triangles. They are also quotients of the modular group *and* they play a key role in another story—the solution of polynomial equations. With this glimpse of wider horizons we invite the reader to explore further, starting with the above books of Gray and Jones and Singerman.

References

M.A. Armstrong
[1979] *Basic Topology*. McGraw-Hill.

E. Artin
[1924] Ein mechanisches System mit quasi-ergodischen Bahnen. *Abh. Math. Sem. Hamburg Uni.* **3**, 170–175.

B. Artmann
[1988] *The Concept of Number*. Ellis Horwood Ltd.

A. Beardon
[1983] *The Geometry of Discrete Groups*. Springer-Verlag.

E. Beltrami
[1865] Risoluzione del problema: "Riportare i punti di una superficie sopra un piano in modo che le linee geodetiche vengano rappresentate da linee rette". *Ann. Mat. pura appl., ser. 1*, **7**, 185–204.

[1868] Saggio di interpretazione della geometria non-euclidea. *Giorn. Mat.* **6**, 284–312.

[1868'] Teoria fondamentale degli spazii di curvatura costante. *Ann. Mat. pura appl., ser. 2*, **2**, 232–255.

M. Berger
[1987] *Geometry I*. Springer-Verlag.

W. Burnside
[1911] *The Theory of Groups of Finite Order*. Dover (1955 reprint).

W.K. Clifford
[1873] Preliminary sketch of biquaternions. *Proc. Lond. Math. Soc.* **IV**, 381–395.

D.E. Cohen
[1989] *Combinatorial Group Theory*. Cambridge University Press.

H. Cohn
[1967] *Conformal Mapping on Riemann Surfaces*. Dover.

R. Dedekind
[1872] *Stetigkeit und die Irrationalzahlen*. English translation (1901) in: *Essays on the Theory of Numbers*. Open Court.

[1877] Schreiben an Herrn Borchardt über die Theorie der elliptischen Modulfunktionen. *J. Reine Angew. Math.* **83**, 265–292.

M. Dehn

[1912] Transformation der Kurven auf zweiseitigen Flächen. *Math. Ann.* **72**, 413–421.

R. Descartes

[1637] *La Géométrie*. English translation (1954): *The Geometry*. Dover.

W. Dyck

[1882] Gruppentheoretischen Studien. *Math. Ann.* **20**, 1–44.

G. Eisenstein

[1847] Beiträge zur Theorie der elliptischen Funktionen. *J. Reine Angew. Math.* **35**, 137–274.

L. Euler

[1748] *Introductio in Analysin Infinitorum I*. English translation (1988): *Introduction to the Analysis of the Infinite: Book I*. Springer-Verlag.

R.D. Feuer

[1971] Torsion free subgroups of triangle groups. *Proc. Amer. Math. Soc.* **30**, 235–240.

L.R. Ford

[1929] *Automorphic Functions*. Chelsea Publishing Co. (1972 reprint).

C.F. Gauss

[1811] Letter to Bessel, 18 December 1811. *Briefwechsel mit F.W. Bessel*. Georg Olms Verlag.

[c. 1819] Die Kugel. *Werke* **8**, 351–356.

J. Gray

[1982] From the history of a simple group. *Math. Intelligencer* **4**, 59–67.

[1986] *Linear Differential Equations and Group Theory from Riemann to Poincaré*. Birkhäuser.

T. Harriot

[1603] Unpublished manuscript, see J.A. Lohne, Essays on Thomas Harriot, *Arch. Hist. Ex. Sci.* **20**, 189–312.

C. Hermite

[1858] Sur la résolution de l'équation du cinquième degré. *Comp. Rend.* **46**, 508–515.

D. Hilbert

[1899] *Grundlagen der Geometrie*. Teubner. English translation (1971): *Foundations of Geometry*. Open Court.

[1901] Über Flächen von konstanter Gaussscher Krümmung. *Trans. Amer. Math. Soc.* **1**, 87–99.

H. Hopf

[1925] Zum Clifford–Kleinschen Raumformproblem. *Math. Ann.* **95**, 313–339.

G.A. Jones and D. Singerman
[1987] *Complex Functions*. Cambridge University Press.

L.H. Kauffman
[1987] *On Knots*. Princeton University Press.

F. Klein
[1872] Vergleichende Betrachtungen über neuere geometrischer Forschungen (Erlanger Programm). *Ges. Math. Abhandl.* **1**, 460–497.

[1876] Über binäre Formen mit linearem Transformation in sich selbst. *Math. Ann.* **9**, 183–208.

[1879] Über die Transformation siebenter Ordnung der elliptischen Funktionen. *Math. Ann.* **14**, 428–471.

[1882] Neue Beiträge zur Riemannschen Funktionentheorie. *Math. Ann.* **21**, 141–218.

P. Koebe
[1907] Zur Uniformisierung der beliebigen analytischer Kurven. *Göttinger Nachrichten*, 191–210.

A.I. Kostrikin and Yu.I. Manin
[1989] *Linear Algebra and Geometry*. Gordon & Breach.

J.L. Lagrange
[1773] Recherches d'Arithmétique. *Nouv. Mém. Acad. Berlin*, Oeuvres III, 695–795.

R.C. Lyndon
[1985] *Groups and Geometry*. Cambridge University Press.

W. Magnus
[1974] *Noneuclidean Tessellations and their Groups*. Academic Press.

W.S. Massey
[1967] *Algebraic Topology: an Introduction*. Harcourt, Brace & World, Inc.

E. Moise
[1977] *Geometric Topology in Dimensions 2 and 3*. Springer-Verlag.

J.M. Montesinos
[1987] *Classical Tessellations and Three-Manifolds*. Springer-Verlag.

J. Nielsen
[1925] Om geodaetiske Linier i lukkede Mangfoldigheder med konstant negativ Krumning. *Mat. Tidskrift B*, 37–44.

V.V. Nikulin and I.R. Shafarevich
[1987] *Geometries and Groups*. Springer-Verlag.

H. Poincaré
[1880] Extrait d'un mémoire inédit de Henri Poincaré sur les fonctions Fuchsiennes. *Acta Math.* **39** (1923), 58–93.

[1882] Théorie des groupes Fuchsiens. *Acta Math.* **1**, 1–62.

[1883] Mémoire sur les groupes Kleinéens. *Acta Math.* **3**, 49–92.

[1887] Les fonctions Fuchsiennes et l'Arithmétic. *J. Math. ser. 4*, **3**, 405–464.

[1892] Sur l'Analysis situs. *Comp. Rend.* **115**, 633–636.

[1895] Analysis situs. *J. École Polytec.* (2) **1**, 1–123.

[1904] Cinquième complément à l'analysis situs. *Rend. Circ. Mat. Palermo* **18**, 45–110.

[1907] Sur l'uniformisation des fonctions analytiques. *Acta Math.* **31**, 1–63.

[1985] *Papers on Fuchsian Functions.* Springer-Verlag.

H. Rademacher
[1983] *Higher Mathematics from an Elementary Point of View.* Birkhäuser.

T. Radó
[1924] Über den Begriff der Riemannsche Fläche. *Acta Univ. Szeged* **2**, 101–121.

K. Reidemeister
[1928] Fundamentalgruppe und Überlagerungsräume. Göttingen Nachrichten, 69–76.

[1932] *Einführung in die kombinatorische Topologie.* Teubner.

G.F.B. Riemann
[1851] Grundlagen für eine allgemeine Theorie der Funktionen einer veränderlichen complexen Grösse. *Werke*, 2nd ed., 3–48. Dover 1953.

[1854] Über die Hypothesen, welche der Geometrie zu Grunde liegen. *Werke*, 2nd ed., 272–287. Dover 1953.

H.A. Schwarz
[1879] Ueber diejenigen algebraischen Gleichungen zwischen zwei veränderlichen Grössen, welche eine Schaar rationaler eindeutig umkehrbarer Transformationen in sich selbst zulassen. *J. Reine Angew. Math.* **87**, 139–145.

C. Series
[1982] Non-euclidean geometry, continued fractions and ergodic theory. *Math. Intelligencer* **4**, 24–31.

J.C. Stillwell
[1989] *Mathematics and Its History.* Springer-Verlag.

H. Weyl
[1913] *Die Idee der Riemannschen Fläche.* Teubner. English translation (1955): *The Concept of a Riemann Surface.* Addison-Wesley.

H. Zieschang, E. Vogt, and H-D. Coldewey
[1980] *Surfaces and Planar Discontinuous Groups.* Springer-Verlag.

Index

Universitext   *(continued)*

Nikulin/Shafarevich: Geometries and Groups
Øksendal: Stochastic Differential Equations
Rees: Notes on Geometry
Reisel: Elementary Theory of Metric Spaces
Rey: Introduction to Robust and Quasi-Robust Statistical Methods
Rickart: Natural Function Algebras
Rotman: Galois Theory
Rybakowski: The Homotopy Index and Partial Differential Equations
Samelson: Notes on Lie Algebras
Smith: Power Series From a Computational Point of View
Smoryński: Logical Number Theory I: An Introduction
Smoryński: Self-Reference and Modal Logic
Stillwell: Geometry of Surfaces
Stroock: An Introduction to the Theory of Large Deviations
Sunder: An Invitation to von Neumann Algebras
Tondeur: Foliations on Riemannian Manifolds
Verhulst: Nonlinear Differential Equations and Dynamical Systems
Zaanen: Continuity, Integration and Fourier Theory